METRIC TALKS

The Second International Conference on Metric Education

Edited by: Marlene M. Milkent

Conference Leaders: John M. Flowers, Executive Director
Lawrence J. Bellipanni, Director
Jack G. Matthews, Co-Director
Marlene M. Milkent, Co-ordinator

Sponsored by: The University of Southern Mississippi
Department of Science Education
College of Science and Technology
College of Education and Psychology

Coordinated by: Department of Conferences and Workshops
Division of Extension and Public Services

Library of Congress Catalog Number: 75-20633
ISBN: 0-8269-4600-3

123456789-75-987654321

PRINTED IN THE UNITED STATES OF AMERICA

CONTENTS

SECTION II—METRIC WORKSHOPS AND PROGRAMS

SECTION III—METRIC-RELATED RESEARCH

PREFACE

The Second International Conference on Metric Education, sponsored by the University of Southern Mississippi, was held January 27–29, 1975, at Biloxi, Mississippi. The conference was attended by approximately 400 individuals, thus making it one of the largest assemblies of metric educators ever to convene. The conference featured a variety of activities, including commercially-sponsored workshops, materials exchange sessions, contributed papers and special interest sessions. All of these were directed toward the common goal of the exchange of ideas among individuals interested in promoting metric education.

The participants of the conference had various educational backgrounds and represented a wide spectrum of special interests. At one end of the spectrum were those who came to learn the basics of the metric system; at the other end were those who have been actively engaged in developing programs, conducting workshops and writing articles related to metric education. Such a diverse group of individuals provided for an atmosphere of active participation which resulted in a vigorous exchange of ideas. While the participants expressed differing opinions and, at times, differing philosophical viewpoints, they were united by a common concern—the pressing need for metric education as a viable part of the American educational system. Among the opinions and recommendations of conference participants and representatives of the special interest group sessions were:

(1) The United States *is on* the metric system, and has been since 1893, when metric standards were adopted as the fundamental weights and measures standards of the United States.

(2) Total conversion to the metric system will involve all phases of business, industry and education. Leaders in these fields are not waiting until Congress acts on the passage of an official metric bill; they are initiating change and exerting influence in order to make metric conversion come about more quickly and proceed more efficiently.
(3) Metric education must begin in the primary grades and must continue throughout the entire educational process. It should be incorporated into all academic areas and it should be taught as *the* language of measurement.
(4) Metric education should not be limited to those engaged in formal schooling. It is imperative that the entire American public be made cognizant of the value of the metric system, informed of its fundamental aspects, and trained to use it.

A general conclusion among conference participants was that individuals must take the initiative in realizing a total conversion to the International System of Units (SI). Educators must pave the way for schools to incorporate metric programs into their existing curricula; business and industrial leaders must prepare their personnel to make the conversion to metric units of measurement. It appears inevitable that Congress will pass legislation mandating that the United States undergo complete conversion to SI. Until then, there is much that can be done to prepare the American public to accept a "new" system of measurement.

The papers contained in this book are representative of the participants of the conference. You will find papers which present differing points of view and sometimes contradictory opinions. It is not the existence of diverse opinions that is surprising but rather the large number of points of agreement among individuals representing various sectors of the educational enterprise. These papers stand as testimony that metric educators possess many common goals and that, by working together, it is possible for them to produce an educational structure which will make the dream of a "Metric America" a reality.

Marlene M. Milkent

LIST OF CONTRIBUTORS

Harold Don Allen, from the perspective of pre-service teacher education, has made a strong contribution to metric awareness and to acceptance of and enthusiasm for SI conversion during the initial years of Canada's metric commitment. His studies in science and education at McGill University, Montreal, were followed by graduate degrees from Rutgers University and the University of Santa Clara. Don Allen has served as elementary school teacher, high school teacher, mathematics department head, elementary and high school principal, and in-service and pre-service teacher educator. Currently on faculty at Nova Scotia Teachers College, Professor Allen recently completed a three-year term as Mathematics-Science Coordinator at the college. Known nationally as a speaker, broadcaster, and columnist on metric education, he is author of *Canada Measures Up* and co-author of *Mathematics with Metric Measure,* a I–VI program based upon the Silver Burdett Mathematics System.

Jim F. Bassett, Ed.D., has taught elementary school mathematics, K–8, in the public schools for the past eight years. Presently, he teaches elementary mathematics methods and supervises student teachers at Sam Houston State University.

Mildred M. Dominy is professor of education and specialist in elementary school mathematics at the State University of New York, College of Arts and Science at Plattsburgh, where she coordinated the Graduate Mathematics Specialization Program. Her graduate work was done at Syracuse University.

Mildred Dominy has served as a mathematics consultant and workshop director in many parts of the United States and has helped numerous school systems in upgrading their mathematics instruction. Professional writings include journal articles, Publisher's Notebooks, co-authorship of the *Harbrace Mathematics* series, teachers' guides for materials published by New Approaches to Learning, Inc., and co-authorship of *Mathematics in Motion,* a book of mathematics activities and games involving total body movement, recently published by Carlton Press. Mildred Dominy is listed in *Who's Who in American Education, Who's Who in American Women* and *Two Thousand Women of Achievement.*

Charles Eicher is an associate professor of elementary education at the University of South Dakota. Dr. Eicher received his B.S. from Otterbien College in Ohio, his M.Ed. from the Ohio State University and his Ed.D. from Indiana University. He served Indiana University as Research Associate and as Assistant Professor of Education before assuming his present position at the University of South Dakota. He has taught elementary school, has been chairman of a junior high school mathematics department, and has served as elementary school principal. Presently, he is responsible for undergraduate and graduate studies in elementary mathematics education at the University.

Clinita Arnsby Ford is currently serving as Director of the Division of Consumer Science and Home Economics at Florida A & M University. Among her many citations and awards are listings in *Who's Who in American Education, Who's Who of American Women,* and *American Men and Women of Science,* the Outstanding Community Service Award from Frontiers International, and the Teacher of the Year Award at Florida A & M University. Dr. Ford has served on numerous committees including the USDA Honors Award Committee, the Role of Women Committee of the American Home Economics Association, and the Program Planning Advisory Committee of the Community Nutrition Institute.

Margaret M. Garr is Mathematics Consultant for Jefferson County Schools at Louisville, Kentucky, acting as a resource person for elementary, middle school and junior high school teachers of mathematics. She has taught elementary school, as well as both chemistry and mathematics at the secondary level. She holds B.S. and M.A. degrees from the University of Louisville. Mrs. Garr has been involved during the past few years in presenting many inservice workshops for mathematics teachers with particular emphasis on metric education.

Anton Glaser was born in a metric country (Germany) and lacks any overzealousness for the metric system that may be associated with new converts. He learned the inch-pound system as a second measurement language after his arrival in the United States in 1939. Dr. Glaser is author of the 1971 *History of Binary and Other Nondecimal Numeration,* the 1974 *Neater by the Meter: An American Guide to the Metric System,* and Penn State's Correspondence Course (available since September, 1974) *The Use and Understanding of the Metric System.* Dr. Glaser is currently Associate Professor of Mathematics at Penn State's Ogontz Campus in Abington, Pennsylvania.

Richard A. Kruppa is an Associate Professor of Industrial Education and Technology at Bowling Green State University, Bowling Green, Ohio. Dr. Kruppa has been active in promoting implementation of SI through journal publications, presentations at national and regional professional meetings, and through workshops conducted at the Bowling Green campus.

Earl C. Leggette is actively involved in teacher education at Jackson State University and has taught a course in metrication for public school teachers using a competency-based approach to instruction. He received his Bachelor degree from Rust College, Masters from Mississippi State University and Doctoral degree from Rutgers University. He is a well-known mathematics instructor. At present, Dr. Leggette is Associate Director of Eight Cycle Teacher Corps and Associate Professor of Secondary Education (mathematics) at Jackson State University, Jackson, Mississippi. His specialty is mathematics education with emphasis on the problem-solving process.

Lottie E. Mackay received her A.B. from Vassar College and Ph.D. from Yale University. Dr. Mackay is presently Senior Science/Mathematics Editor for Schloat Productions, Tarrytown, New York. In the past (interspersed with raising four children), she has done research in organic chemistry for Standard Oil Development Corp. (now Exxon Research) and for Burroughs-Wellcome Co. and has served as Science Consultant for Herman Kahn of the Hudson Institute; she was also Science/Mathematics Editor for Educational Audiovisual Co.

Jean Hunter McKibben received her B.A. in home economics from Hiram College, Hiram, Ohio. A former vocational home economics teacher, Mrs. McKibben is presently a member of the American Home Economics Association and the New Mexico Home Economics Association; President of the New Mexico Home Economics Association in 1972–74 and State Counselor in 1974–76; Chairman of the American Home Economics Association Metric System Committee for 1975–76; Chairman of the New Mexico Home Economics Association Metric Committee and State Metric Coordinator; Member of the U.S. Metric Association, Inc., American National Metric Council, New Mexicans for Metrication, and New Mexico Academy of the Science Metric Committee.

Marlene M. Milkent received her B.S. in Education from California State College (Pennsylvania) and her Ph.D. in Science Education from the University of Texas at Austin. Dr. Milkent is currently Assistant Professor of Science Education at the University of Southern Mississippi. She has been actively involved in teaching the metric system to elementary education majors as part of their pre-service training. She is interested in all areas of science education from the elementary to the graduate level, and as such engages in a variety of activities including developing science courses for nonscience and elementary education majors, conducting in-service workshops for elementary teachers, and supervising graduate students in science education.

Gomer Pound is Dean of the Division of Extension and Public Service at the University of Southern Mississippi. He has studied at Michigan State University and Florida State

University, earning his Ph.D. in Music Education from the latter institution. Prior to joining the faculty at the University of Southern Mississippi, Dr. Pound taught at Morehead (Kentucky) State University and Florida State University. A member of several professional and honorary societies, he is listed in *Leaders in Education* and *Outstanding Educators in America.* Dr. Pound has written many articles that have appeared in professional publications, has been a contributor to *Music Educational Materials,* and has authored a textbook, *Music Fundamentals for Teachers.*

Willard F. Reese is a professor of science education at the University of Alberta, Edmonton, Canada. In addition to many journal articles and television and radio programs, Dr. Reese has developed several metric teaching aids. His game "Travel Metre", designed for active participation with metric units and to acquaint the players with international road signs while sharing an enjoyable experience, is commercially available. For further information please write to Dr. Reese at the University of Alberta.

Ivan W. Roark received his B.S. in Education from the University of Missouri and is a Certified Manufacturing Engineer, Society of Manufacturing Engineers. He has industrial experience, including Douglas Aircraft, and has taught in technical areas in industry. Professor Roark has been active for 20 years in the Society of Manufacturing Engineers. After more than 23 years experience, he is presently Professor Emeritus at the University of Tulsa and has been the Coordinator of Technology at Tulsa Junior College since its inception in 1969.

Armand M. Seguin received his B.S. degree from St. Cloud State College in 1966, his M.S. from Indiana State University in 1967, and his Ed.D. in Industrial Education from Arizona State University in 1974. Dr. Seguin has taught electronics technology and is currently the Coordinator of Graduate Studies for Industrial Arts at Jackson State University. He has been active in metrication by directing workshops, writing, giving lectures, consulting, making television appearances and as an active member of the Mississippi Metrication Committee.

Robert L. Taylor received his B.S. and M.A. from the University of Texas, and his Ph.D. from Colorado State University. He is a retired Lt. Col., U.S. Air Force. Dr. Taylor is presently Instructor of Mathematics and Science, Community College of Denver, North Campus.

Robert M. Todd is Associate Professor of Mathematics and Education at Virginia Polytechnic Institute and State University where he works in teacher training. His current interests include measurement, metrication and mathematical learning disabilities. He has earned degrees in civil engineering, education and mathematics at Purdue University and his Ed.D. at the University of Virginia.

John H. Trent is a Professor of Mathematics and Science Education at the University of Nevada, Reno, Nevada. Previously Dr. Trent served as an Associate Professor of Science and Mathematics for Ohio University in Athens, Ohio, and Kano, Nigeria. His experience includes 15 years of teaching high school mathematics and science. He has written numerous articles relating to environmental science and mathematics education and has directed several National Science Foundation (NSF) institutes in these areas.

Larry J. Weber is Professor of Education at Virginia Polytechnic Institute and State University. A member of the Psychological Foundations Program area, Dr. Weber's specialization is educational measurement. He has worked extensively in the field of education program evaluation and student evaluation by teachers.

EMERGING PERSPECTIVES ON EDUCATION FOR A METRIC WORLD

Harold Don Allen

The palm trees, even after the Hurricane Camille battering, look pretty good in January. The Gulf and the outdoor pool are not really cold—only the natives think so! I'm delighted to be back with you and able to share metric education experiences here on the Mississippi Gulf Coast. To give credit where due, the Air Canada ticket agent at Halifax International Airport, Nova Scotia, combined her undoubted talents with the airline's *megadollar* computer facility to liberate me from our hard-packed snow and get me *here,* all in one day. I'm grateful. The alphabet soup of international air travel quite fascinates me—I've taught combinatorics. *AC 673* (that's Air Canada's early morning flight) was to jet me from *YHZ* (Halifax) to *BOS* (Logan International at Boston); *EA 145* to *ATL* (that's Eastern to Atlanta); then *SO 517* (Southern), with remarkable hopscotch to Alabama, Florida, and Mississippi. "There," my Air Canada girl said, double-checking the *GPT* computer readout. "You're confirmed right through to Gulfport . . . *wherever that is!*"

"Wherever that is"—meaningless symbols from computer output, a three-letter permutation without the vision of a reality of battered palm trees, millionaires' yachts, and shirtsleeves weather in January fog. I think that there is a message here. I can see a real risk, an understandable one, that we on occasion err in somewhat this direction in our work in metric education. All of us, I think, are rightly conscious of our basic SI—*Le Système International d'Unités,* our current, international metric. We know, for example, the *newton* as the unit of force. Those with a physical sciences background may cite the mass-

Harold Don Allen, F.C.C.T Nova Scotia Teachers College, Truro, Nova Scotia, Canada

acceleration definition. But what really *is* a newton? What would you pick up—a pingpong ball, a telephone receiver, a typewriter, or *what*—to experience a force (in this context, a weight) of *one newton?*[1] This kind of knowledge or insight I fear we may lack. Such a lack must not go unremedied. Textbook SI is absolutely essential, but in itself is not enough. It's *necessary* but not *sufficient,* to allow my mathematical training to show through. We may recite *attometre, femtometre, picometre, nanometre,* all the way through *giga-* and *tera-* (the whole correct structure of currently adopted prefixes), and yet not know that a person's *height* (and waist measure and clothing sizes) are properly given in *centimetres;* and, I think worse, not be able to give our own and our family's heights, to the centimetre, without fumbling with attempts at conversion. We need a feeling, a strong, intuitive feeling, for units and measures in the real world—one's mass in kilograms (to the kilogram), a size 60 waist measure, a five kilometre walk, one hundred millilitres of corn syrup, a two hundred degree oven (in degrees Celsius), a fifty kilometres per hour road speed, a sensible tire pressure (in kilopascals). A publication of the (United States) National Council of Teachers of Mathematics has made reference to a *500 gram* vitamin pill.[2] Figuratively and literally hard to swallow, it'd have the mass of an average loaf of bread. *Milligrams,* of course! This sort of error shouldn't happen. When working apart from the context of a metric reality, it so easily can.

Equally, we could err in the opposite direction—preach schoolboy metric as perhaps we learned it: *ergs, dynes, grams force;* CGS, MKS, MKSA, a smattering of some or all; obsolete symbolism (gm, dkm, Km); unconventional groupings (the *myriametre*—do you recall?); non-decimal subdivisions (one-third of a litre); variant spellings . . . as distinct from *Système International,* to which our world, and I trust, we, are committed. Here, I assert, there is no defense. Ignorance (as with the law) can be no excuse. Before you teach or preach a metric term in *1975,* you are duty bound to check yourself out on the modern, internationally developed, decimally based *Système International,* the "world metric" into which French metric of the 1790's has been evolved. Your students are to inherit a predominantly SI world. Right now, from their teachers, their school texts, their learning aids, their measuring instruments, North American children are entitled to nothing less than metric that is SI—pure and simple.

I think it highly desirable that we view SI in proper perspective. There is nothing "foreign" about it, certainly not in the sense that the United States or Canada will be shifting to another nation's measures. No nation is yet on pure SI. SI is a world aspiration. France and Germany, for example, are committed to shift to SI, which should require of them some very fundamental and far-reaching rethinking. The *kilogram force,* for instance, must be basic to the mental processes of the European laborer—yet the *newton* (named for an Englishman!) and not the kilogram is the force unit (the *only* force unit) in SI. SI is the international metric of the International Organization for Standardization (ISO). It is rooted in recent decisions of General Conferences of Weights and Measures (CGPM's) and of the International Committee of Weights and Measures (CIPM), to which leading nations had effective input. The "most SI" country today, perhaps significantly, is not a traditionally "metric" country; on the contrary, most would concede it to be the Republic of South Africa, a nation which has legislated a rapid, efficient conversion to metric measures that are specifically SI. While the United Kingdom, Australia, New Zealand, Canada, and most other nations that now are in the process of SI conversion, have opted for a voluntary changeover (that is, have chosen persuasion rather than compulsion), as no doubt will the United States, a great deal can be learned from South African experience to date, as reported in an attractive diversity of specialized publications and newsletters.[3]

Teachers have spoken to me of the difficulties which they and their pupils encountered in the "metric" chapter of the traditional mathematics text. I think back to long tabulations of "metric words" (many never really used), to exercises calling for little more than meaningless "symbol shoving," to measurement words devoid of association with measurement experiences, and I don't wonder the trouble and distaste. We know and have demonstrated in abundantly practical terms that metric education, especially knowledgeable SI metric, needn't be anything like that.

SI, I am convinced, will simplify teaching of measurement, thinking about measurement, and measurement itself. Simplicity will stem from three characteristics consciously built into SI.

1. SI is decimal. Like the numeration system which it can serve to reinforce, SI groups by tens. SI subdivides into

tenths. Computation becomes easier when decimals to appropriate precision replace a diversity of nondecimal fractions.

2. SI is coherent. Units themselves are simply related. Remember the *horsepower?* In schoolboy physics, a horsepower represented a rate of doing work of *550* foot-pounds per second. Note the factor, *550.* In SI, power is measured in *watts.* A watt is the power that produces energy at the rate of *one* joule (newton-metre) per second. The "one" reflects coherence.
3. Each SI unit is unique and serves one purpose. Pressure, for example, is measured by the *pascal.* An incredible diversity of pressure units (many of them "metric") will eventually come to be replaced. Consider a few: *atmospheres, pounds per square inch, bars* and *millibars, millimetres of mercury, dynes per square centimetre.* One SI unit (the pascal) replaces all. Each unit serves one purpose. Our greatest confusion in measurement concepts stems from the fact that *pound* (and kilogram!) have traditionally been used to measure both mass and force.

SI is decimal. As the language and thought patterns of measurement undergo their inevitable shift to "metric" and SI, implications of this decimal structure become apparent. A metre is divided not into *halves, quarters,* and *eighths* (watch your learning aids!) but into *tenths, hundredths,* perhaps *thousandths.* Precision takes on marked importance in this context. "One-third of a metre" is 33 cm, if one folds cloth (measured to the centimetre), but 333 mm if sheet metal (conventionally measured to the millimetre). Common fractions all but vanish from measurement language. Collaborating on a Grade I–VI mathematics textbook series for Canadian schools, I had this truth brought home to me when I was left with the assignment of providing "word problems" for operations with fractions.[4] When you've done the obvious thing about one-third of a pizza plus another one-sixth of a pizza, where do you get your problems? I'll tell you where. You cheat a little, turn to the few remaining vestiges of non-decimal measurement, and do "$3\frac{4}{7}$ weeks plus $2\frac{3}{7}$ weeks" or "$5\frac{1}{2}$ hours minus $2\frac{3}{4}$ hours"! Fractions remain highly significant as ratios and as rational numbers, of course, but their role in measurement (which has been decreasing for a century—think decimal feet, decimal degrees!) is rapidly approaching the vanishing point.

In this increased stress on decimals and decimal measures (and corresponding decreased emphasis on non-decimal fractions and their manipulation) I see two implications:

1. Earlier introduction of decimal notation and earlier teaching—with understanding—of *tenths* subdivision, should be feasible. There is close identification among "two tens and three ones," "two dimes and three cents," and "two decimetres and three centimetres." Correspondingly, "six metres, seven decimetres, eight centimetres, nine millimetres"—alternately written 6.789 m—takes the student to three places of decimals, potentially far earlier than we have seen fit heretofore.
2. Reading specialists speak knowingly of "readiness." My experience as principal of a school which had five-year-olds in Grade I assures me that they have a point. Do we find a valid analogy in the computational skills of school mathematics? Is there "fraction readiness," a stage at which the child has command of the underlying operations and motivation to face the drudgery that many associate with fraction manipulation in Grades V and VI? Perhaps we have worked for too much too soon. Properties of the positive rationals have great significance in mathematical structure, but not necessarily in the "real world" of our ten-year-old. Possibly unit fractions and simple ratio comparisons can suffice until structure is to be studied for its own sake. I realize that properties of rationals are needed even to solve a linear equation with natural coefficients. Not division of mixed numerals and such, however. Identifying a common denominator—*any* common denominator—and "multiplying through" can eliminate most computational hurdles.

The real educational challenge of an SI world, to me, is to teach not metric but *measurement* as never before. In old textbooks (I collect them) the format of the mensuration section was remarkably predictable. Units were tabulated (inches, feet, yards, rods, furlongs, miles), with appropriate equivalents. Available for reference, the table might or might not be committed to memory. One suspects that it was! Exercises called for appropriate calculation. No one ever measured a rod or walked a furlong, at least not on school time! Formulas for areas, volumes, etc., of certain regular figures were presented or developed. Problems called for the "plugging in" of appropriate values in the formulas, then computation. Little if any emphasis was placed on precision or significant digits. Measurement, I am convinced, is to be learned primarily if not solely by experience, and not from a book. Measurement is not a chapter or topic, but rather a continuing theme. It should begin with early and meticulously correct presentation of fundamental concepts, and broaden and deepen as the learner grows in maturity and sophistication. When standard units are SI and conversion factors are powers of ten—essentially decimal shifts—rather than 12's, 16's, 36's, 1760's, etc., the emphasis is off computation as such, and hopefully on such underlying themes as estimation, accuracy, precision, and true comprehension.

SI is a language of measurement, as it stands (it continues to evolve) the most elegant measurement language yet to be devised by man. Like any language, it is best learned by immersion. Look to a total school experience that is metric, for optimal learning. You may feel (most Americans do) that you cannot justify elimination of traditional units, at least not yet. I'm not sure I agree. Let me offer you some comfort, however, in the form of a Canadian anecdote. The late Louis S. St. Laurent, Canada's postwar prime minister who lived into this decade as a highly respected elder statesman, once was asked by a reporter his views on aspects of Canadian bilingualism and biculturalism. He replied, in effect: "When I was a boy, I did not know that there were two Canadian languages, just one with which I spoke to my father and one with which I spoke to my mother." My point: two languages, even two measurement languages, *can* be learned side by side. The child can grasp, in parallel development, centimetres, decimetres, and metres, and inches, feet, and yards. Just don't "convert"! One must justify, of course, time spent on two systems, one clearly destined to be of decreasing importance. For the younger pupil, who will graduate into a markedly metric country and world, I cannot justify such a course of action.

Our students' great need and our teachers' great challenge will be (in unattractive but appropriate psychological jargon) to develop a "gut feeling" for SI *as it applies to the real world.* This will call for a strong, intuitive feeling for the units. It will call, too, for insight into emerging world practices and a certain amount of second guessing as to the practical role of SI units in all sectors of an increasingly metric North America. I caution, in this connection, against undue emphasis on the *kilometre-hectometre-decametre-metre* type of unit substitution, so beloved of traditional texts. What the student will want and need is a good working acquaintance with those units most likely to be encountered in daily living. Let us consider several common types of measurements.

Linear measure (length, width, height, perimeter, thickness) will see four SI units dominate in daily experience. The *kilometre* typically will measure distances between towns. To me a kilometre is a ten to twelve minute stroll. The *metre* will measure a foot race, a building lot, or a golfer's eighteenth hole. Use of the *centimetre* will be restricted, in the main, to body measures (height, arm length, bust, waist, hips—a "perfect 36" becomes 90-60-90, with sensible rounding) and cloth and clothing measure. The precision of most shop work will call for the *millimetre,* as will, typically, a geometry diagram. One other unit, convenient and desirable in connection with a number of areas of curriculum, is the *decimetre.* It is a good unit for work with young children, a logical multiple of the centimetre or subdivision of the metre, and a stepping stone to the definition of the litre.

Three—only *three*—mass units cover a broad range of daily experiences. The *kilogram,* the base unit, will measure sugar, flour, potatoes, luggage—or the student's own body mass. The *gram* will suffice for most smaller quantities. Postal rates no doubt will advance in twenty gram steps. The megagram, the metric ton or *tonne,* will measure carload lots. *Gram, kilogram, tonne* will handle a wide range of measurement needs. The *milligram* and *microgram* may serve, as now, to measure very tiny allotments—for example, the additives in so-called "enriched" foods.

The square hectometre of SI—an area unit—has associated with it the special name of *hectare.* It is likely to retain its world importance as a unit of land measure. Other common area units will be the *square centimetre* (or centimetre squared), *square metre,* and *square kilometre.* The hectare is ten thousand square metres (10 000 m^2) or one hundredth of a square kilometre (0.01 km^2). Correspondingly, the cubic decimetre of SI—a volume unit—has associated with it the special name of *litre.* In common practice the litre is a unit of fluid capacity. Two related measures are the *millilitre,* counterpart of the cubic centimetre, and the *kilolitre,* counterpart of the cubic metre.

The familiar *second,* SI unit of time, has sexagesimal and other non-decimal multiples, harkening back to Babylonian numeration, but is subdivided decimally. Resulting units are the *millisecond, microsecond,* and—notably in computer science—the *nanosecond.*

Typical SI units of school science, the *newton, pascal, joule, watt, ampere, volt,* etc., form their multiples and submultiples in terms of powers of one thousand, all of which are legitimate units. Questions of appropriateness resolve themselves to the SI rule that a unit should be so selected that, in general, a numerical value in the range 0.1 to 1000 will result.

What will be the role of what Americans call "customary" units in a dominantly SI future? For one thing, to fly over the non-metric checkerboard of the American midwest is to see how old units and practices persist close to the soil. Pre-revolutionary French units still exist in Quebec and Spanish units in Texas and California. Canadian government maps, even now, however, superimpose on a diversity of distinctive land-use patterns a uniform square kilometre grid. Use of non-SI will ever be a function of time and place. Non-Olympic sports may retain traces of old measures, much as a horse is measured today in *hands* and raced in *furlongs.* The day will come, however, when "pound" and "mile" are taught, like the Biblical *cubit,* as literary and historic references—as, for example, Shakespeare's "pound of flesh."

Who are the potential leaders in the vital task of bringing the elegance of SI to a school community? I often am asked this by teachers who feel that it's someone other than themselves! The science teacher and the mathematics teacher have a lengthy involvement with metric units, but their record of bringing these into the real world is far from impressive. The science lab somehow has been a place set apart. Further, the conventionally trained (FPS, CGS, MKS, MKSA) science teacher has more metric relearning and rethinking to do than anyone else on faculty. Until he bones up on SI, he's a menace, I'd say. The mathematics teacher should be in a position to produce impressive results within his area of competence, but he first must rethink his whole teaching of measurement. Teachers in those curriculum areas where measurement skills have been most carefully and most successfully taught—I refer to the industrial arts, home economics, aspects of physical education—should have a distinctive contribution to make. North American Olympics in 1976 should focus new attention on world records—which, of course, are metric. The most effective of school metric advocates may turn out to be the student leaders, at least that's my experience. In student government, school newspapers, cocurricular activities, their influence may accomplish what faculty alone cannot. For optimal learning, the school itself should strive to be an SI world. From track and field to shop to social studies to math lab, the challenge *can* be met—and at some point should.

SI is a living language. As such, it cannot be mastered—by you or your students—in an ivory tower. One must live with it, think in it, work with it. Only in this way can be picked up the fluency essential for optimal teaching and learning. Know your SI—*know it cold*—and develop a strong feeling for how it is used and how it will be used. These references, among others, should be helpful:

National Standard of Canada: Metric Practice Guide, from Canadian Standards Association, 178 Rexdale Blvd., Rexdale, Ontario.

Metrication: A Guide for Consumers, from Canada Department of Consumer and Corporate Affairs, Ottawa, Ontario.

Metric Conversion & You, from Australia Metric Conversion Board, P.O. Box 587, Crows Nest, N.S.W.

The Use of SI in Primary Education, The Use of SI in Secondary Education, Metrication for the Family, and *Precision Conversion of Dimensions,* all from South African Bureau of Standards, Private Bag 191, Pretoria, Republic of South Africa.

Conversion implies change, and SI conversion will call for a startling diversity of changes. Measurement permeates all of organized living. The most subtle, most difficult, changes are those involving people's minds. Interaction with concerned people—all kinds of people—provides a wealth of insight. Student teachers from my classes got a liberal education by manning a Metric Information Booth at a local shopping centre.[5] My favourite communications medium is "open line" radio, the phone-in talk shows where the elderly, in particular, will listen to what I as guest have to say, then will call in with their apprehensions. As educators, we vitally need this contact, this exploration of SI implications in human terms. We can allay fears, most of them groundless, and at the same time identify genuine areas for concern. In everything we do as educators, we need and work best with public support. In helping to ready our society for an SI changeover, we can marshall such sound arguments that strong backing should be assured.

Metric conversion represents, in purely economic terms, a perfect illustration of (in the particularly apt phrase of the United States Metric Study), a decision whose time has come.[6] There is more to all of this, however, than long-term economic gain. In SI—strict SI—I see for all of us a new freedom. This freedom is three-fold:

1. Freedom from a heritage of needless numerical manipulation, awkward factors being eliminated or reduced to a minimum by the decimal nature and coherence of SI.
2. Freedom from ambiguities and uncertainties—pound mass or pound force, American gallon or Imperial?—in a system that has a single, unique unit for each measure.
3. Freedom from idiosyncracies and inconsistencies in a measurement system which establishes standards for the world.

I sense in our meetings and in my conversations here in Mississippi an undercurrent of feeling that there may be something that goes against individual freedom in the adoption and enforcement of national and international standards of measurement. I see this as nonsense, but I acknowledge that you may have to think yourselves through. Perhaps I can help with an illustration which, while near trivial, does for us come strikingly close to home. When our youngest was born last July, she was large by any neonatal standards. Our children were suitably impressed. Yet, in truth, she wasn't as large as our oversized family cat, not quite. Today, she is. We have, consciously or otherwise, compared her size to the cat as she has grown—gained mass—these past six months. To do so among ourselves, while not overly scientific, is our privilege—if you would, our freedom. When we communicate with others, however, we find it fitting to refer to a 4.6 kg birth mass (that's how the birth announcement read), to 7.5 kg at six months, or to a

5.0 kg cat. It is good to have standards that can be universally understood. Like *horsepower* (a picturesque unit), *catmass* has outlived its particular usefulness, at least for us.

I leave you then to your good work, with this injunction: to learn SI, to translate it into terms of real life, and to consider carefully and fully what it implies and will imply for your curriculum and your students. The sun is shining, and I intend to go out into this 16°C Mississippi January and stroll a few kilometres up the beach. Yes, I really do think that way. You should, too. A few kilometres (lower case km) toward Gulfport *(GPT).* The time is past when a metric educator should say or even feel, "Whatever that is!" or "Wherever that is!"

CHAPTER NOTES

1. Two golf balls are about right, although my personal preference is a preselected 102 g rock—the force (weight) on the palm is $m \cdot a = (0.102 \text{ kg})(9.8 \text{ m/s}^2) = 1.00 \text{ kg} \cdot \text{m/s}^2 = 1.00 \text{ N}$.

2. C. William Engel, "Are You Ready to Go Metric?," *The Mathematics Student,* 21:1 (October, 1973), pp. 1–3.

3. In particular, *South African Metrication News* (monthly), available without charge by surface mail from South African Bureau of Standards, Private Bag 191, Pretoria, Republic of South Africa.

4. Sidney A. Lindstedt, Harold Don Allen, and Iris Schickler, *Mathematics with Metric Measure,* I–VI. Based upon the Silver Burdett Mathematics System of the United States, this Canadianized program is published by GLC Educational Materials and Services, Ltd., 115 Nugget Avenue, Agincourt, Ontario.

5. Accounts of a number of our metric education activities remain accessible. See, for example, community cablecasting depicted in *Metric News,* 2:1 (1974): 7, and a variety of teacher education experiences, reported in "Our Clock Says 13 . . . and it's Time to Talk Metric," *Kappa Delta Pi Record,* 10:1 (1973): 4–5.

6. United States Metric Study, *A Metric America: A Decision Whose Time Has Come* (Special Publication 345, 1971; Washington: United States Government Printing Office, 1971), available at $2.25.

TO TEACH THE METRIC SYSTEM—ISOLATE OR INTEGRATE?

Lottie E. Mackay

Needless to say, the question asked in the title is rhetorical. *Of course* we integrate the metric system into all parts of the curriculum. And *of course* we realize that includes not only science and mathematics, but every subject area at every level, wherever any kind of measurement is used or discussed. So the real question becomes, not *whether,* but *how* we can best make the metric system a part of the whole curriculum.

Before we try to answer that, let us first look at how the metric system gets into *any* part of the curriculum. Who, in other words, is responsible for metric education in your school? In your district? In your state? There are many different answers, the most usual one being the *mathematics supervisor* or *department head.* In some instances it is the *science specialist.* There are also examples of cooperative ventures between *mathematics* and *science specialists.* And some states and districts have designated a *metric coordinator* whose prime responsibility is metric implementation.

The chief concern of a metric coordinator (using this now as a generic term) is to implement metrication in his area of jurisdiction, usually by including metrics in the mathematics and/or science syllabus. Such inclusion may range all the way from completely mandatory directives (with mandated timelines), through recommended guidelines and timelines, to completely discretionary suggestions.

But there are a few prerequisites for the success of even a mild program of metrics in the curriculum. Teachers must be properly trained in all relevant aspects of metrics, and feel suf-

---

Lottie E. Mackay, Ph.D. Senior Science/Mathematics Editor, Schloat Productions, Tarrytown, New York.

ficiently at ease with metric units to pass their knowledge on to their students. This in turn requires the availability of appropriate teaching materials, including textbooks, audiovisual materials, and measuring tools—both for teacher training and for actual teaching. Finally, the ease and success with which a school system can go metric will depend in large measure on the favorable attitude of the public in general and of the district's parents in particular.

Needs of this kind were addressed, among others, by last year's Interstate Consortium on Metric Education[1], which made recommendations on metric education on behalf of 27 adoption states and territories. Although these 27 do not speak for the entire country, ICME's guidelines represent the most concerted and best organized effort to date to bring about some kind of uniform action in metric education. Another, less detailed, set of guidelines for teaching measurement comes from the National Council of Teachers of Mathematics (NCTM) in the form of an article, "Metric: not *if* but *how,*" written by members of NCTM's Metric Implementation Committee.[2] The two sets of guidelines are substantially in agreement, and should be of considerable help to metric coordinators across the country.

So far we are still, for the most part, talking about metrics in the mathematics and science curriculum. How do we get "the whole student" involved? How can we get a student population really "thinking metric" rather than just using the metric system in certain specified situations?

Let us begin to find answers where it is easiest, and also most logical: at the elementary level. Right away we have two things going for us: the students are still largely unfamiliar with *any* measuring system, so they can learn to use the metric system as they learn to measure. And the teacher with whom they will learn about measuring and metrics is the same person who will be teaching them social studies, language arts, and most of whatever else they will learn in school. Certainly there is plenty of opportunity to bring in some aspect of the metric system in discussions of history, geography, economics, and even language arts. And this is in addition to the actual *use* of metric units in the course of normal classroom work. For example, the teacher who is using metres in 4th grade math problems would surely rather refer, in social studies, to a coast-to-coast distance of 4 800 kilometres than of 3 000 miles. In other words, the elementary teacher, who teaches the whole student, can be readily convinced that the student is better off dealing with only one system of measurement.

In junior and senior high school the situation gets a little more difficult as teaching becomes fragmented into separate subject areas taught by different teachers. So the "metric people" must be willing to do a certain amount of missionary work not only with their own mathematics and science teachers (who are likely, at the very least, to need some updating of their knowledge of old-style metrics to modernized SI metrics), but with teachers and supervisors in the social sciences and humanities. This will not be easy, and will take some planning and organizing. Lectures and workshops to familiarize teachers with metric units will be needed, as well as constructive suggestions on how historical, economic, geographical, political, and philosophical aspects of the metric system are relevant to the existing curriculum. Certainly, too, there are a great many possibilities for independent research and enrichment projects that can be suggested to non-mathematics/science-oriented teachers for their students. Examples might include:

History of man's need to measure and how he has coped with it.

Economic pros and cons of U.S. metrication.

U.S. metrication—how it affects and is affected by national and international politics.

Systems of measurement as means of communication.

Many other useful ideas and suggestions can be found in education magazines.[3]

But enrichment projects are not limited to social studies and humanities. Mathematics and science themselves offer some interesting possibilities that fall outside their normal curricula. Some examples are:

Devise an ideal measuring system, defining base units in any way you choose; explain reasons for your choices.

Devise an ideal numeration system (e.g., base 12), and use it to formulate an ideal measuring system; explain reasons.

Suggest possible natural replacements (present and future) for the prototype kilogram as the base unit of mass in the metric system. Discuss any problems associated with your suggestions.

Does any all-encompassing internally consistent measuring system require 7, and only 7 independent base units? Explain.

Discuss possible methods and consequences of metricating time.

So far I have given only passing mention to some very important fields that are vitally concerned with metrication: home economics, industrial arts, vocational education, and sports. Interest in metrics from people in these fields may in the long run do more for getting public acceptance of the metric system than all the zeal that mathematics and science can muster. Specialists and students in each of these areas are vitally concerned with measurements, and must face not only the coming of the metric system but also the hard facts of life here and now in a partially metric world. For example, the vocational student learning to repair automobile engines gets first-hand experience with the realities of metric and non-metric engine parts, each requiring different sets of tools.

It is precisely because educators in these fields have been confronted with such problems that many of them realize that metrics are coming, and that metrics are easier. In fact, far from having to be pushed and cajoled by their local "metric coordinators," many of them are their school's or their district's prime metric activists. Cooperation between these specialists and the metric coordinator can open up all kinds of avenues to "total metrication" in your school or school system.

I hope I have given some constructive suggestions on ways to broaden the metric base in your school or school system. The task is big, but the opportunities are many. Let us use them in ways that make "going metric" an interesting pursuit for the whole school population.

CHAPTER NOTES

1. This consortium was sponsored by the Mathematics Education Task Force of the California State Department of Education, and was funded by the U.S. Office of Education. Its purpose was to improve leadership in the State Education Departments in the field of metric education.

2. This article first appeared in the *Arithmetic Teacher* in May, 1974. It was reprinted in NCTM's *A Metric Handbook for Teachers.*

3. Two recent examples are: S. Beckman and J. D. Hunt, "Metric Mapping," *Science Activities* (November/December 1974): 30–31; R. K. Sparkes, "Metrics is Also a Social Study," *Teacher* (October 1974): 56–62.

DISCREPANCIES BETWEEN PLANNED AND ACTUAL USE OF THE METRIC SYSTEM

Anton Glaser

"Take 1 Kilo of Spinach . . ." was the title of a recent newspaper column by James A. Beard. Of course, the plan called for "kilogram" instead of just "kilo." Passages in foreign-language literature sometimes refer to "10 degrees cold" when −10°C is meant. These are two examples of discrepancies between plan and practice.

Many European reference books admit that hectometer and dekameter, though permissible, have failed to catch on. People have politely declined to use them. Here is an example where planners assumed that every prefix would be "popular"—but only certain ones turned out to be so.

We might distinguish between two kinds of discrepancies. In the one kind ("kilo" for "kilogram") usage is "wrong." In the other, usage is correct, but unexpected—such as the unexpected usage that shuts out deka and hectometer. Both kinds, however, might be blamed, at least in part, on the planners. Is not "kilo" (when kilogram is meant) a mild form of rebellion against the overly long names of metric units? The non-use of deka and hecto is an indication that the original planners provided too many choices.

Of course the *plan* has not remained static. Indeed some of the revisions may be traced back to discrepancies between earlier plans and practice.

SI is the latest plan. It wisely narrows down choices among previously permissible forms. Surprisingly it declares radians the angular measuring unit. Do the planners remember the "grad" (1/100 of a right angle) that failed to catch on? Does

---

Anton Glaser, D.Ed. Pennsylvania State University, Ogontz Campus, Abington, Pennsylvania.

the radian have a better chance? Will the physicians give up measuring blood pressure in millimeters of mercury and replace it by pascals?

Most European countries have written laws that require SI units to be phased in by January 1, 1978. Will they meet their deadline, or will there be postponements and exceptions reminiscent of receding deadlines for certain pollution control equipment in cars? It is important to note that SI is still largely a plan.

Let us return to the question of prefixes and implications for metric education in the U.S. We shall make reference to three sets of metric education guidelines in this connection, namely·

(1) NCTM (National Council of Teachers of Mathematics) published in the May, 1974, issue of *The Arithmetic Teacher.*
(2) AIT (Agency for Instructional Television) *Metric Education* September, 1974.
(3) NBS (National Bureau of Standards) Guidelines *implicit* in its single-page flyer *All You Will Need to Know About Metric (For Your Everyday Life)*

Fortunately, all three are in agreement that for elementary school and everyday life the following metric units suffice (for length, mass, and volume):

mm, cm, m, km

g, kg

ml, l

Nevertheless it is easy to find instructional material which, contrary to these guidelines, gives hecto and deka equal billing with milli and centi. This can be compared with giving furlongs, rods, and leagues equal billing with inches and feet.

SI can be described as a small subset of previously permissible units of measurement. Did the SI planners consider the consumer and practitioner? Of course, yes—but perhaps not as much as they should have. The physician, who may be asked to switch from millimeters of mercury to pascals, may ask, "What is to be gained?" The "coherence" of SI is an advantage when one computes, but the physician merely measures the blood pressure, interprets its meaning for the patient, but does not do any computation with the measured value.

What do cm^2 and cm^3 mean? How do you pronounce them easily? In algebra ab^2 means abb and ab^3 means abbb. Does cm^3 mean 1/100 of a cubic meter? Mr. Richard Deming, author of *Metric Power* seems to think so, for he announces that

$$1\ cm^3 = 0.01\ m^3.$$

This is off by a factor of 10 000. One reason the metric system is called a decimal system is one tends to make errors that are powers of ten.

Since exponents seem to pose great difficulties for a large part of the population, one would hope that ISO (or some other appropriate international body) would consider some alternate symbol for cm^3, perhaps even cc.

Should the consumer and practitioner have greater influence on future decisions concerning SI? Yes, unless planners enjoy their position as scolders of practitioners who won't follow the plan. One can imagine little Johnnie coming to school crying:

TEACHER: "Why are you crying?"

JOHNNIE: "My father is in jail. They arrested him because they found a kilo of heroin on him.

TEACHER: "I'm really upset, Johnnie. If I've told you once, I've told you a thousand times. Don't say "kilo" when you mean kilogram."

METRICS Á LA CARTE

Mildred M. Dominy

As our country moves from the customary to metric (SI) system of measurement, every teacher is faced with *three* immediate concerns:

1. How will changing to the metric system of measurement affect what must be taught?
2. What will be required of me, the teacher?
3. How do I go about teaching that which needs to be taught?

Many elementary/junior high school teachers are just getting over the trauma brought about by "Modern Math." Those having a number of years of experience in the classroom still remember the fear and trepidation of being told that "tomorrow" or "next week" they must teach the "new math" . . . perhaps District "X" or School "Y" was already teaching it—and of course, one MUST keep up with the new and be ready to jump on the bandwagon! The outcomes of this approach were many; and as might be expected, some were good and some not so good. Actually, the predicament of the teacher was not brought about by "Modern Math," but by the manner in which it was being implemented.

Let's review or examine some of the facets involved. Major problems included such things as:

1. "Modern Math" or "new math" meant different things to different people, e.g.:

Mildred M. Dominy, Ph.D. Professor of Education, State University of New York College of Arts & Sciences, Plattsburgh, New York.

a) what one group called "new," another might call traditional, depending on the program or approach previously used by a teacher or school system;
b) some interpreted it to mean that their current program should be tossed out and replaced by the "new," or science of number approach. . . . with much attention given to sets and set theory, systems of numeration other than ours, mathematical structure, algebraic expressions, geometry, etc.

2. Teachers, as well as parents, didn't always understand WHY the mathematics programs were changing. Often teachers, and even curriculum coordinators, had little or no opportunity to share in planning the changes—and the "pure mathematician" was calling the shots, not the mathematics educator who not only had knowledge of the subject but also "know-how" insofar as teaching and learning it.
3. Many teachers felt that they could not teach the "new math," or they felt uncomfortable with it as they had far too little understanding of the new ideas and terminology themselves. To help, in-service programs were organized, either by the school district or in conjunction with the State Education Department or a local college or university. In some cases attendance was required—or volunteered under pressure—which didn't help teacher attitude. Effectiveness of these programs varied considerably. Some were really helpful, but all too often a secondary mathematics teacher or a college mathematics professor was hired to conduct the sessions, and they did so by lecturing on mathematics content. The language and concepts were often geared above the basic needs of the elementary/junior high school teacher not having a strong mathematics background, thus leaving them more confused, or with a more negative attitude, than they had before. In such situations ideas and/or techniques to be used in teaching elementary students were minimized or, if included, were often inappropriate.
4. Many pupils were penalized, mathematically, as they became guinea pigs for the innovators, experimenters, and the pure mathematics enthusiasts.

Now, in retrospect, some of the errors made are obvious, as were the attempts at distortion of the basic goals of the elementary/junior high mathematics curriculum. By virtue of world events, e.g., Sputnik, the mathematics education pendulum was pushed too far off-center. In the rush to be different, and to correct all past wrongs, many would-be educators went overboard—just as had happened in the past. They forgot momentarily that there are *three* important factors to be considered in the development of any curriculum—the subject, yes, but also the needs of the student and of the society in which he exists. "New Math," by some interpretations, attempted to give priority to the subject, which reverts back to the days of the Romans when the Patricians of the day spent their leisure hours discussing the "science of number."

Now that the smoke has cleared and we have had an opportunity to reflect and evaluate, a better balance has been achieved, and when we speak of "Modern Math" we know that we really mean finding better and more meaningful ways of teaching the important basic mathematical concepts and understandings with use of more exact terminology. Greater emphasis is placed on thinking, seeing relationships, and using them; and less emphasis is placed on sheer memorization. We realize, more than ever before, that mathematical ideas evolve, often intuitively, and grow and expand with experience and maturity—that there are levels of learning and understanding. All persons will NOT become mathematicians, just as all will not wear the same size coat or have the same likes and/or dislikes; but hopefully, all will become mathematically literate to the extent required by our society or to achieve a desired career opportunity or goal. Even this is a big job if you think of the actual time available. Did you ever stop to think that typically the student receives only about 80 to 100 hours of mathematics instruction per year (i.e., 30–40 minutes per day—with approximately 180 school days per year)? It is imperative, therefore, that teachers/curriculum coordinators determine HOW to use this limited amount of time wisely and to best advantage by being selective as to what should be taught—and how. It is not what mathematics CAN you teach, but what OUGHT you to teach.

You are probably wondering why I am re-hashing mathematics curricular problems at this time, but I do feel that there is an analogy insofar as the situations the teacher will face during the conversion to metric measurement. Without careful planning, many teachers and students will be in for more of the same. Yes, those of us here are sold on the advantages of the metric system, and want the change to come about. Many others recognize that such a change could be of value, but surveys show that the majority of the "general public" have not yet accepted the fact. Last summer at the beginning of a Metric Education Workshop which I offered on my campus, a check as to the attitudes of the participants toward changing to metric was made. Now these individuals were, with two or three exceptions, teachers who were there because they recognized the need to be prepared to use and teach the SI measurement. Over half of the group were *personally* against changing—and many of the reasons given sounded very like those expressed by teachers 10 to 15 years ago when they were resisting "Modern Math". . . . . . though it was of interest to note the "converts" by the end of the workshop experience . . . . . and to hear the usual "I just didn't realize . . . . ." We know that many teachers have not faced up to the truth that they will have an important role to play in facilitating the change, that they have an obligation to their students, and that they will be held accountable. Many are even resisting self-help via workshops, in-service courses and the like. They are waiting until they have no choice.

For one activity, the summer group mentioned earlier turned its attention to "sampling" the sentiments of the general business/industrial/political population of the area. A survey form, adapted from the one developed by the B. F. Goodrich Company and reported in the May/June 1974 issue of *Metric News,* was prepared. Each class member then arranged for interviews with randomly selected persons, using the form as a basis for gathering and recording information. The individuals contacted were leaders or representatives of small industrial and business firms, area schools, city, county, state, or federal personnel. The results of these interviews were revealing, but not surprising or unexpected. Basically they were as follows:

1. Most first learned about the metric system—or heard of it—at the junior high level. A number remembered converting from °F to °C in science, etc.
2. Few admitted having much knowledge of the system, and a number who felt they had a "moderate knowl-

edge" demonstrated during the interview that they really had very little.

3. Most were aware of the contemplated change, but didn't feel that it would really affect them to any degree; others were vague about the whole thing and doubted that it would really happen.
4. Priority given to planning for change or implementation was very low in most cases—there was a "sit and wait" attitude. One quote from the warden of a well-known correctional facility bluntly expressed what many did not put into words. He stated that he knew absolutely nothing about the metric system; he had received no information concerning metrication plans for his facility; and that until orders came from the State Department, nothing metric would be initiated. An interesting aside was that the warden's secretary stopped the interviewer on the way out and said that she was 60+ years old; even so, she thought people around there should be thinking more about going metric!
5. Actually, few seemed to think that a great expenditure would be required if, and when, they did change.
6. About half of the people interviewed thought some type of educational, in-service, or on-the-job training program would be helpful *if* it were geared to what they would really need and be required to use.
7. Favorable responses to the change to metric were received most often from Air Force personnel (especially those having had an overseas assignment); people working with auto parts, water pollution control, and pharmaceutical products (interestingly enough, not hospital employees); manufacturers of snow-mobiles and heavy equipment; and Canadian firms with subsidiary plants in the area, e.g., Phentex.
8. The one statement often repeated was, "Can't see any reason to change, but when we have to, I suppose we will."

Let's see where all this leads those of us interested in metric education. Michael Nolan has said that "Happiness is best attained by learning to live each day by itself. The worries are mostly about yesterday and tomorrow." Maybe this is a good recipe for an orderly transition to metric—it is useless to worry about what we have not done; but we can learn from past mistakes. If we will give the "now" all we have, and work each day to achieve the outcomes we desire, there will not be too many metric worries in the future.

1. It is vital that we as educators *do* something "in the now" and become active in word and deed, and believe, a la Oral Roberts, that "something good is going to happen."
2. Plans should be made for a gradual, non-threatening transition; and, just as for Affirmative Action, such plans should be on file and reviewed periodically. Someone (or group) should be held accountable. If it takes a little pressure, e.g., State Education Department to schools, State Department to state-supported facilities, company heads to departments, etc., well, there is no time like the present. In fact, with present conditions (recession, etc.), it may be the ideal time to devote thought to change and be ready to start implementation.
3. Let's provide the adult population with opportunities to become accustomed to thinking metric. This can be done through parent groups, Scout leaders, non-credit courses, J.C.E.O. programs, county or state agencies, social/welfare workers, agricultural or homemaker groups, adult education opportunities at schools, facilities, senior citizen centers, etc.
4. State Education Departments, colleges, and school systems should become more active in "picking up the ball" insofar as in-service training for teachers and training teacher consultants. The teacher should have the training *before* being mandated to teach metric. But, let's not just have in-service programs to have them. Guidelines should be developed. The sessions should provide background and knowledge of the metric system (SI) AND a developmental learning sequence for teaching it from kindergarten on—with appropriate help for selection of and use of teaching aids and materials at every level. And, let's have people who know what teaching is all about conducting these programs. Oftentimes the teacher can learn best by using the same laboratory, hands-on approach as that provided

children. If the teacher learns, and has fun doing so, we can expect positive action resulting in pupils being helped, not hindered, in use of the system. As the teachers gain understanding, many opportunities will present themselves to the alert instructors. The metric system with its decimal notation, the patterns, and the relationships, relates so perfectly to the Base 10 system we use.

5. The more recent mathematics and science materials developed for use at all levels have taken the plunge and provide at least a limited number of experiences using metric measurement. Teachers should be encouraged to NOT omit these lessons, but instead, to enlarge upon them and omit some of the work using the customary system.
6. Now, I come to what may pose a problem, and the reason for the title METRICS À LA CARTE. One of the reasons teachers are afraid of the change to metric is that they are overcome by the many ramifications—the do's and don'ts of conversion. They hear the language, see all the symbols, hear of the "seven basic units," base units, supplementary units, derived units, multiples of units, and they are trying to sort out which relate to space and time, periodic and related phenomena, mechanics, heat, electricity and magnetism, light, acoustics, physical chemistry and molecular physics. Many of the references are to technical uses and the teachers are already lost without a familiar frame of reference for even everyday products, sizes, dimensions, area, volume, etc. Many articles refer to industrial change related to fasteners, diameter/pitch combinations, metric threads, plugs, gages and the like. Oh, yes, then there are new paper sizes—and even the correct way of writing numerals is changing!

At this point the teacher is overwhelmed and, often, "turned off." The unknown can be terrifying! All of this is foolish, of course, but it won't go away and we must appreciate these concerns and attempt to alleviate them through a common-sense approach. After all, we do not teach everything about using our customary system, and we were never expected to do so. The job of the school is to provide the student with the basic

> knowledge adequate for living in our society and all the specialized needs are provided later, as the situation demands, e.g., career or vocational training, on-the-job training, etc. In short, if a parallel is to be drawn, only the organization and structure of the metric system and the metric units which will be required and used need be taught at the elementary/junior high school levels. Emphasis will be on the basic SI units for length and mass and units derived from them. There must be familiarity with the few non-SI units used extensively in commerce, e.g., hectare, litre, metric ton, kilometres per hour, kilopascals for tire pressure and the Celsius temperature scale. Correct writing and reading of commonly used symbols will be necessary.

It behooves the teacher then to select from the "Metric Menu" that which is appropriate, desirable and pertinent to members of our society. This is why it is not enough to "teach metric" *per se,* but there must be a sequentially developed teacher's guide, a curricular plan for metric education, and it must be followed.

If teachers are participating in adult training, or in business or industrial on-the-job programs, here again, appropriate choices—and different ones, perhaps, must be made from the menu. A final caution might be: Use common sense, don't try to force-feed, and don't overwhelm!

THE HOME ECONOMIST STIRS METRICS INTO ADULT EDUCATION

Jean McKibben

Since 1866, use of the metric system in the United States has been legal, but not mandatory. Despite several attempts to convert the country, Americans have stubbornly clung to a measuring system that was already outmoded over a century ago. In 1968, Congress authorized the Secretary of the Department of Commerce to conduct a study of the advantages and disadvantages of finally adopting the metric system as the American way. In August, 1971, a report, which was based on a study conducted by the National Bureau of Standards, was sent to Congress by the Department of Commerce. In response to the recommendations of this report, a number of bills were introduced into both houses of Congress. To date, no legislation has been passed. Nonetheless, I think all of us agree "inevitable" is the adjective for metrication of the United States.

The American Home Economics Association (AHEA) is one of the few professional organizations actively concerned with promoting metrication. In 1967, the Association passed a resolution in support of measures that would promote adoption of the metric system. In 1970, the Association co-sponsored with the Bureau of Standards a Consumer Conference which provided some of the information used in the Department of Commerce's report to Congress. The Association's committee responsible for the American National Standards Institute Standard on Dimensions, Tolerances, and Terminology for Home Cooking and Baking Utensils has a subcommittee on metrication. That subcommittee is responsible for making recommendations on the size and metric name for the tablespoon and

Jean Hunter McKibben, B.A. Member, American Home Economics Association and State Counselor, New Mexico Home Economics Association.

teaspoon and for the metric markings on the cup. It will also recommend sizes in metric dimensions for cooking and baking utensils.

AHEA also has asked that each of its state organizations select a person to coordinate information on metric conversion. I serve New Mexico in that capacity.

For the last four years, the metric system has been a special project for the Home Economist in Homemaking section of AHEA. Home economists in this section have been informing and teaching members of their state associations and local groups through prepared papers, talks, workshops, and displays on the metric system.

Many think a home economist does nothing but "stitch and stir." Whether you realize it or not, nearly every day brings you in contact with a home economist or the result of her work, for home economists work in education, business, extension, food and nutrition, family economics, home management, consumer interests, health and welfare, textiles and clothing, housing and household equipment, institutional management, and communications. With this range of interest, home economists have a great opportunity to exert their influence and expertise during conversion to the metric way of life. Many home economists in these various fields are already preparing for conversion by presenting papers, programs, and workshops to their fellow workers and to the consumers with whom they come in contact.

Moreover, since the essence of home economics is teaching the public how to live better, the home economist is uniquely qualified to answer the call of Ms. Valerie Antoine, Metric Association Vice-President and Metric Planning Consultant to industry and educators, who pleads, "What is needed now is to speed up the public's understanding and acceptance of what is inevitably in store—a metric way of life for all Americans."[1]

The American home economists are prepared to do this and are now ready to stir metrics into adult education. Our recipe for "Metric Conversion" has only three ingredients: one, a home economist enthusiastic about the metric system or a teacher, preferably a home economics teacher, trained to teach the metric system; two, a group of consumers; three, a plan, program, workshop, or display.

This recipe, like any recipe, requires a certain amount of equipment before it can be assembled. The equipment necessary for displays and workshops or classes is:

Volume—Dual-marked cups, or $\frac{1}{4}$, $\frac{1}{2}$, and 1 litre vessels and standard 1, 2, and 4 cup measures for comparison; and measuring spoons in metric size.

Length—Meter stick, tape measure, rulers, and linear scales.

Mass—Dual-marked scales, both bathroom and kitchen.

Temperature—Dual-marked thermometer.

Workbook—Simple charts, exercises, bibliography.

Unfortunately, most of this equipment, unlike educational aids for other teachers, is not readily available. It must be ferreted out in small, out-of-the-way, specialty shops, or ordered by mail in massive quantities. In either case, a considerable investment of both time and money is required. This is a real disadvantage, for most of us are volunteers with no financial backing. We therefore rely heavily on free educational packets occasionally offered by various companies. (J. C. Penny's has a good one called "Moving Toward Metrics.")

The kitchen equipment I have for my display was purchased mostly from import-gourmet shops. My tables for converting recipes were ordered all the way from Africa. Of the three cookbooks in my display, two come from Africa. I have been able to find only one metric cookbook published in the United States, and that one is entitled *American Cooking for Foreign Lands.*[2] Everywhere I go I search for cookbooks, measuring equipment, shop tools, anything metric since I keep hoping metric items will appear on convenient, local store shelves.

The difficulty of getting equipment easily and inexpensively undermines all our efforts to convince the public that conversion need not be confusing and expensive. The consumer interested enough to attend a class or workshop quickly loses his budding enthusiasm when he discovers he can't get the equipment he needs at his local store and that he has to pay a great deal for it. He goes away saying, "When the equipment is available, I'll start to work with it." Meanwhile, he'll talk against conversion to others. It is our hope that each person attending a workshop or class will talk enthusiastically about metrics and thereby interest others.

Once we have enough equipment, we can begin to assemble our recipe for "Metric Conversion." Place the three ingredients in a meeting room, classroom, consumer workshop, educational fair or display, and stir. There are many ways to mix together a teacher, a group of consumers, and a program. The method will vary according to the size and make-up of the group, the time allotted, the location of the meeting, and the equipment available. I will give you an outline of the way I've been presenting and teaching the SI system.

I always start with the basic terms. Usually I open the meeting or workshop with Dr. Flowers' film, *The Survey of Modernized Metric.*[3] Here the consumer hears the basic terms defined and learns the symbols and correct spelling for the metric units.

Then I present the SI Metric system as a separate system, using as few conversions as possible. I use a few conversion tables to illustrate how much easier the SI Metric system is. SI Metric can also mean "SImple Metric." Nothing scares an adult more than tables and charts listing measurements in four decimal places.

I teach metric units as a separate language. I urge my participants not to practice converting from one system to another. When using dual-marked tools and equipment, they are to compare the two systems briefly, but to use the metric side. I suggest they mark cookbooks, recipes, diagrams, etc., in metric units. Then they are to forget the old and use the new. At this point in the program, I distribute the Bureau of Standards' chart, "All You Need to Know About Metric for Your Everyday Life." I ask them to speak a little metric everyday and to look around their homes for examples of metric already in use.

With the metric terms explained, the participants are ready to learn metric by doing. I have everyone present using metric units as they play games and measure various objects. As one student said, "If I hear, I forget; if I see, I remember; if I do, I understand." We all learn by repetition.

During this period of learning, I explain the use of the equipment and tools I have on display. I show them what equipment they'll need for cooking and sewing and tell them where it can be obtained. Most dress patterns are now dual-marked. Metric tape measures, hem markers, rulers are all easy to get. Dual-marked liquid measuring cups in $\frac{1}{4}$, $\frac{1}{2}$, and 1 litre sizes are beginning to appear in local stores.

Metric kitchen and bathroom scales can be found. Kitchen scales, however, are not necessary. Although I suggest that some might like to try preparing their favorite recipes by means of metric scales (since weighing dry ingredients is a more accurate method in any system), I do not emphasize it. Because most Americans are unfamiliar with weighing dry ingredients on kitchen scales, the home economists coordinating metrication encourage them to continue to measure their dry ingredients by volume. Asking an American cook to use metric kitchen scales is asking him to break two long-held habits. The simpler we keep the change-over, the quicker we will get the public to participate and accept conversion.

While the participants in my workshops and classes are playing games and measuring everyday objects, they express various personal feelings about changing to a metric way of life. Some common ones point out some problems we must overcome. One that particularly shocked me came from not one, but several home economics teachers. They said with vehemence, "I'll teach it, but I won't use it." This is utterly ridiculous. To teach the metric system, one must be enthusiastic and creative enough to inspire your pupils to explore metrics and to become proficient in using the new system. Today's pupils are tomorrow's consumers.

About the first comment ever made is a complaint about having to throw away all of one's cookbooks and cooking utensils. To counter this, I explain that no one has to replace everything at once. A cook can continue to use her old system cookbooks until they fall apart. All that is necessary is to convert to metric units the recipes used and to use dual-marked measures, which have been acquired gradually. Or, simply use old utensils with old cookbooks and new metric utensils with new metric cookbooks. Eventually only old-fashioned grandmothers will cook with cups and teaspoons. Young modern cooks will use 250 ml and 5 ml measures only.

Many homemakers say they can live the rest of their lives without metrics. True. Likewise, they can live the rest of their lives without TV, a car, or electricity. But who wants to? Let's keep up with the rest of the world. I urge them to read about the SI Metric system and recommend Frank Donovan's book, *Prepare Now for a Metric Future.*[4] I ask them to use metric terms and tools more and more; think metric; become enthusiastic. When you dream in metric, you're ready for tomorrow.

If more of the public would approach the metric system as several senior citizens did, our job would be so much more rewarding. Instead of seeing problems, these people were delighted to see something made fun and easy. "Oh, I can do that," was their comment about the metric way of measuring.

Most of the negative reaction from the public is due simply to a lack of instruction on how the metric system will affect everyday life. Industry, large businesses, and most of the professions are beginning to conduct workshops and training sessions to help their employees deal with the metric system in their work. No one, however, is helping the consumer, small businessman, and homemaker to deal with it in their daily living. As a result, they see problems, but no answers.

For example, a fabric shop owner was completely bewildered over how he could convert since he didn't know how to figure prices in metric and, in fact, didn't know one thing about the SI Metric system. As I explained to him, the fabric companies will mark his bolts or material pieces in meters; the clerk will read on the measuring meter or stick meters and centimeters instead of yards and fractions of yards; and the cost will be figured according to meters.

Likewise, a shoe clerk complained he couldn't fit shoes with all those crazy sizes. As I explained to him, the shoe manufacturers will determine the sizes as they do today; when the shoe fits, put it on. Similarly, to the consumer who was afraid she couldn't find her size when she didn't know anything about metric sizes, I explained that she need only to try on until she finds the size she wants and then she need remember only that one number, for in a SI Metric world sizes will be standardized and thus won't vary from company to company or even from country to country. To the woman who already knows her metric size, but is horrified at its large number, I remind her that while her size may be larger, her mass will be reduced by over half.

As these reactions illustrate, the public is confused and scared. When metrication is mentioned, it sees nothing but problems. The answers to these problems must be provided. This can be done through a network of adult workshops and small classes, where the individual can find the answers to his individual problems and worries and, at the same time, where he can learn the basic facts that he'll need to function in a metric America. Once each individual understands how metrication will affect him in his daily life, in other words, once the public is educated, conversion will be at hand. For, an educated consumer will demand the proper legislation, will insist that metric tools and equipment be on his local store shelves, will use metric units every day. He will be the most powerful tool we can use to get conversion accomplished thoroughly and quickly.

This is a job the home economist is prepared to do. Although it is a tremendous job, many of us are already working hard to prepare other home economists and other interested persons to teach and inform the public. We are now ready to stir metrics into adult education by taking one enthusiastic home economist, one group of consumers, and one program and then mixing well. "Metric Conversion" will soon be done.

CHAPTER NOTES

1. Evelyn De Wolfe Nadel, "Getting Used to the Metric System," unknown airlines magazine (Spring 1974), p. 21.

2. Maj-Greth Wegener, *American Cooking for Foreign Lands* (Greenwich, Connecticut: North Castle Books, 1969).

3. John M. Flowers, Ed.D., *Filmstrip 1. Survey of Modernized Metric* (Wichita: Library Filmstrip Center, 1973).

4. Frank Donovan, Prepare Now for a Metric Future (New York; Weybright and Talley, 1970).

IMPORTANCE OF COHERENCY IN THE INTERNATIONAL SYSTEM OF UNITS

Ivan W. Roark

ABSTRACT

To the non-public the term *coherency* of the International System on Units (SI) provokes a suggestion of definitive rhetoric. It is the purpose of this paper to expand the definition sufficiently to establish the proper concept in usage.

DIMENSIONS

Since the beginning of time man has struggled to maintain himself within the framework of his environmental existence, his personal attributes clashing with fundamental qualities and quantities of natural phenomena. The scope of his analytical activities is divided into many abstract components such as fear, hunger, anger, hate, love, pity, sympathy, happiness, pain and many others which he has accepted, each with his own concept of magnitude. How excellent it would be to have a measure, a standard, by which these abstractions, in their intensity, could be conveyed to one another. In a recent article a physician expresses dismay because he "did not know the level of the threshold of pain." Only one of these abstractions, love, has a measure. The Great Teacher, the Messiah, said, "No man hath greater love than this, that he lay down his life for his neighbor"; but there seem to be no sub-multiples to the extreme. These fundamental qualities may well be defined as DIMENSIONS with which we cope within the framework of our existence. We lack units within the dimensions by which to express both quantity or intensity.

Ivan W. Roark, Coordinator of Technology, Tulsa Junior College, Tulsa, Oklahoma.

Better progress has been made with physical phenomena within man's environment. He has always been confronted with dimensions of time, length, mass, heat and cold (temperature), light, sound, and their derivatives, some including combinations involving several dimensions. Perhaps the earliest sense developed in man was a sense of proportionality—enough of this, enough of that—to satisfy his system requirements. From this unorganized beginning many of the traditional systems of measurement have been developed, all of which lack stability in proportionality even to this day. In our present system, the inch system, we find many special names for quantities which give no indication of the dimensions from which they are derived, if, indeed, they were so derived. Much confusion exists in expressions of volume, mass, density, force, weight, and other dimensions indicating an amount, enough of a substance, to effect a comparison. For example: we have 2 miles, 4 tons, 2 pounds, 2 ounces, 2 systems of volume and an innumerable number (or kinds) of bushels. As the old saw goes, "a bushel of wheat, a bushel of rye, a bushel of clover. . . ." even a bushel of coal. Volume and mass are confused.

Special names for multiples and sub-divisions of units within our dimension of length give no indication of the unit from which they are derived; neither do they have a common constant of proportionality. Traditionally inches are divided into binary fractions: $\frac{1}{2}$, $\frac{1}{4}$, $\frac{1}{8}$, $\frac{1}{16}$, $\frac{1}{32}$, $\frac{1}{64}$. A few multiple units are:

4 inches = 1 hand
9 inches = 1 span
12 inches = 1 foot
3 feet = 1 yard
$5\frac{1}{2}$ yards = 1 rod, pole, perch
40 poles = 1 furlong
8 furlongs = 1 mile
3 miles = 1 league

A paradigm for reference is necessary to determine the conversion factors. The above are examples of current divisions in the dimension of length; other dimensions are similarly constituted, each having a total lack of homogeneity within its own structure and a lack of coherence within the measuring system which requires the use of conversion constants when derivations are made. No thought is given to the convenience of the decimal system. ANSI B87.1-1965 does specify decimal divisions of the inch but the use of the traditional system prevails!

PROPORTIONALITY

A more tenable system has been accomplished in the SI system of selecting units which have a one-to-one proportionality (1:1). Dimensions which are most used by engineers are surprisingly few in number; length, mass, time, temperature and electric current. Luminous intensity (candela (cd)) and quantity of substance (mol) make up the seven SI units but they are little used except in their specific fields of engineering. Their concept is little realized by the general public except in the form of applied rationalization. The one-to-one proportionality between the unit dimensions is known as COHERENCY. The real significance of coherency is attained by the simplicity of calculations in SI units; no conversion factors are needed. They are eliminated by the constant of proportionality 1:1.

COHERENCY

In a search for a concept of coherency it is well to examine a few definitions which are found in current literature relevant to the use of SI units.

ASTM Standard Metric Practice Guide E 380-70

> The original metric system provided a *coherent* set of units for the measurement of length, area, volume, capacity and mass based on two fundamental units: the metre and the kilogram. Measurement of additional quantities required for science and commerce has necessitated development of additional fundamental and derived units. Numerous other systems based on these two metric units have been used. A unit of time was added to produce the centimetre-gram-second (cgs) system adopted in 1881 by the International Congress of Electricity. About 1900, practical measurements in metric units began to be based on the metre-kilogram-second (MKS) system . . . . . in 1950 the ampere, the unit of electrical current, was established as a basic unit to form the MKSA system.
>
> The great advantage of SI is that there is one and only one unit to each physical quantity—the metre for length (ℓ), kilogram (instead of gram) for mass (m), second for time (t), etc. From these elemental units, units for all mechanical quantities are derived. These derived units are

defined by simple equations such as $v = d\ell/dt$ (velocity), $a = dv/dt$ (acceleration), $F = ma$ (force).
Note: cgs is not a coherent system since a centimetre has a conversion factor of 1/100 metre and the gram 1/1000 kilogram.

IBM SI Reference Manual

The *coherent* nature of SI is preserved by defining all derived algebraic combinations in terms of unity, thereby eliminating conversion factors within the system. For example, the derived unit of power, given the special name *Watt,* is defined as one joule of work done in one second of time.

Metrification for Engineers by Ernst Wolf
A publication of THE SOCIETY OF MANUFACTURING ENGINEERS

A *coherent* system is defined as a system of units in which the product or quotient of any two unit quantities is a unit of the resulting quantity. The coherency in SI has the following corollary conveniences:
1. The same system and unit of measurement is used regardless of discipline, industry or trade involved.
2. A *minimum* of conversion factors other than powers of ten are needed.

Basic Training Guide to the New Metrics and SI Units by Robert C. Sellers
NATIONAL TOOL, DIE AND PRECISION MACHINING ASSOCIATION

SI is designed to be coherent. The units are chosen so that when a compound quantity is made up of any combination of the six primary units, each having a value of one, the compound quantity itself has a value of one. For example:

The coherent unit of force is the newton. One newton is one kilogram metre per sec. per sec. A non-coherent unit of force is the dyne. One dyne is one gram centimetre per *sec. per sec., i.e., (10^{-3} kilogram) × (10^{-2} metres) per sec. per sec. or 10^{-5} kilogram metres per sec. per sec.*

SOCIETY OF MANUFACTURING ENGINEERS TECHNICAL PAPER MM74-892
Importance of Correct Metric Usage by
Louis F. Sokol, President, US METRIC ASSOCIATION

It (SI) is a unique system in which the product or quotient of any two unit quantities is the unit of the resulting quantity. This characteristic simplifies engineering computations where the factor is generally one. The following relationships show this coherence factor:
FORCE = MASS × ACCELERATION 1 N = 1 kg · m/s^2
WORK, ENERGY = FORCE × DISTANCE 1 J = 1 N · m
POWER = ENERGY/TIME 1 W = 1 J/s

Terms previously used to describe the proportional nature of the current metric systems include such terms as: proportionality, constant of proportionality, consistency and homogeneity, all of which impress the non-scientific mind with a measure of rhetoric. The cgs system is said to be consistent, which means, perhaps, that since one centimetre is .01 m, an accepted sub-multiple of a metre, and one gram is .001 kg, an accepted sub-multiple of 1 kg there is a consistent set of proportionalities in the system. Only the SI system fully defines coherency and makes optimum use of it. As may be seen from the above definitions coherency is relevant only to those who make computations when the system is put to use. The following examples may be sufficient to generate an understanding of coherency. But first we must have a few definitions.

SI has but one unit for each dimension.

SI is a decimal system.

Multiple units are whole numbers.

Sub-multiple units are expressed as decimal fractions of a unit.

Derived quantities (derivations) may be made up of multiples and sub-multiples of one unit or they may consist of operations involving both multiples and sub-multiples of several unit dimensions.

Examples

Dimension: Metre	Symbol m (note small m)
one metre = 1 m	Unit dimension
1 km = 1000 × m	A multiple of basic unit
Area = $\ell \times \ell = \ell^2$	A derived quantity

By substitution of the unit of length *m*

$$1\text{ m} \times 1\text{ m} = 1\text{ m}^2 = \text{Area}$$

To find the area of a rectangle:

15m

$\ell \times \ell = \ell^2 = A$ 5m

Fig. 1

computation

$\ell \times \ell = \ell^2 = \text{Area}$

$15\text{m} \times 5\text{m} = 75\text{ m}^2$

It may be written:

$(15 \times 5) \times (\text{m} \times \text{m}) = 75\text{ m}^2$

Figure 1 shows that the system is coherent: the lengths are expressed in the basic unit m.

In Figure 2 it is required to express the area in units of square inches and, also, square feet. By name, both one foot and one inch are recognized as units of length in the traditional inch system. To find the area in square inches a proportional constant must be found to convert feet to inches. This will make a coherent system having 1 inch as the unit of length.

2 feet

8 in. $\ell \times \ell = \ell^2 = A$

Fig. 2

24 inches

$A = 192\text{ in.}^2$ 8 in.

Fig. 3

In figure 2, $\ell \times \ell = \ell^2$, area in square units. But the system is not coherent. To make the system coherent it is necessary to find a constant of proportionality for converting the unit lengths to the same basic unit dimension. Thus: to convert 2 ft. to inches:

1 foot = 12 in.; divide by 1 ft.:

$$\frac{\cancel{1\text{ foot}}}{\cancel{1\text{ foot}}} = \frac{12\text{ in.}}{1\text{ foot}} = 1\text{, the constant of proportionality.}$$

Then in inches $= 2\not{ft.} \times \frac{12 \text{ in.}}{1\ \not{foot}} = 2 \times 12 \text{ in.} = 24 \text{ in.}$

By substitution in Figure 3 the system is coherent. Then,

$$24 \text{ in.} \times 8 \text{ in.} = 192 \text{ in.}^2 \text{ or } (24 \times 8) \times (\text{in.} \times \text{in.}) = 192 \text{ in.}^2$$

To find the area in square feet a conversion factor can be established and calculations made in terms of unit feet, or inches can be considered as a sub-multiple of a foot. Thus:

$$12 \text{ in.} \times 12 \text{ in.} = 144 \text{ in.}^2 = 1 \text{ sq. ft.}$$
$$144 \text{ in.}^2 = 1 \text{ ft.}^2$$

Divide by 144 in.^{2}

$$\frac{\not{144\ in.^2}}{\not{144\ in.^2}} = \frac{1 \text{ ft.}^2}{144 \text{ in.}^2} = \text{C (constant)}$$

Then: area, in square inches $\times$ C, equals area in square feet.

$$192\ \not{in^2} \times \frac{1 \text{ ft.}^2}{144\ \not{in^2}} \qquad \frac{192}{144} \times \text{ft.}^2 = 1.33 \text{ ft.}^2$$

These calculations, usually considered quite superficial, are actual operations needed to express the areas. Coherency, as a name, may be new to some but the principle involved is of long duration. Conversions from one basic unit to either a multiple or a sub-multiple as a base will involve similar calculations. There is a singularity in the use of the unit dimension in SI that, due to the use of powers of ten, makes the system desirable. The following paradigm is included to show the advantage of coherent units in derivations in SI.

Quantity	*Dimension*	*Calculation*	*Derivative*
Area	Square Metres	$1 \text{ m} \times 1 \text{ m} =$	1 m^2
Volume	Cubic Metres	$1 \text{ m} \times 1 \text{ m} \times 1 \text{m} =$	1 m^3
Velocity	Metres per Sec.	$1 \text{ m} \div 1 \text{ s} =$	$1 \text{ m}/ 1 \text{ s}$
Distance	Velocity $\times$ Sec.	$1\text{m} \div 1\text{s} \cdot 1\text{s} =$	$1\text{m}/1\text{s} \times 1 \text{ s}$
Pressure	Newton per m^2	$1 \text{ N} \div 1 \text{ m}^2 =$	$1\text{N}/1\text{m}^2$
Watt	Volts $\times$ Amps	$1 \text{ V} \times 1 \text{ A} =$	$1 \text{ VA} = 1$ Watt
Force	Newton	$1 \text{ kg.} \times 1\text{m}^2 \div \text{s}^2 =$	$1 \text{ kg} \cdot \text{m}^2/1\text{s}^2$

Thus, calculations can be made by use of coefficients of units without involving constants of proportion.
Example:

How far does one travel in 30 sec. at a speed of 90 kmh?
$1 \text{ hr} = 60 \times 60 \times \text{sec.}$ $\qquad 1 \text{ km} = 10^3 \text{ m}$

Statement: $\frac{90 \times 10^3 \times 30 \text{ sec.}}{60 \times 60 \times \text{sec.}} = 750$ m, or using parallel

phases, $\frac{\overset{3}{\cancel{90}} \cdot 10^3 \cdot \cancel{30}}{\underset{2}{\cancel{60}} \cdot \underset{2}{\cancel{60}}} \times \frac{m \cancel{s}}{\cancel{s}} = 750\text{m}$

RATIONALIZATION

In the SI system, conversion factors for multiples and submultiples of units are expressed in powers of ten. However, there are certain physical constants which do not meet the requirement of coherency but because of their specific properties are accepted by the SI. Such physical constants are much used in calculations and their usage must be applied as constants other than unity. A few of these are listed below:

$\pi = 3.1416$
Speed of light = 299.8 Mm/s
Speed of sound = 331 m/s

Time is not true SI because the conversion factor, 60, is not a multiple of 10; only the second is SI. The unit of circumference, the degree, is not SI: it is to be expressed in radians. Kilometres per hour is not SI; only m/s. There are many physical constants used by engineers and scientists which are not coherent but due to their special nature there is no alternative.

There are some areas which present a conflict with SI because of their traditional usage. Miles per hour most assuredly will become kilometres per hour. One litre was to be one dm^3 (cubic decimetre) but the usage of litre has been long established as the one probable metric unit most of the people know. These are areas of rationalization. Many areas of rationalization will appear as the USA goes Metric. What will be the unit of pressure for automobile tires? What will be the unit of barometric pressure? How many "cubes" of butter per kilogram? Will we purchase in terms of mass or newton? The list is long and yet to be established. It is the contention of the author that the metric system cannot be properly taught until we can know the areas of rationalization which will be applied to SI by industry as its usage increases.

SOURCES OF MATERIAL

International Business Machines Corp., *SI Metric, Reference Manual,* White Plains, N.Y., I.B.M., 1973.

Wolf, Ernst, *Metrication for Engineers,* Dearborn, Michigan, Society of Manufacturing Engineers, 1974.

Sellers, Robert C., *Basic Training Guide to the New Metrics and SI Units,* Washington, D.C., National Tool, Die and Precision Machining Association, n.d.

Sokol, Louis F., "Importance of Correct Metric Usage,"*Technical Paper MM74-892*, Dearborn, Michigan, Society of Manufacturing Engineer.

American Society for Testing and Materials, *ASTM METRIC PRACTICE GUIDE (A Guide to the Use of SI-International System of Units),* Philadelphia, PA, ASTM, 1972.

International Standards Organization, *ISO Standard 1000-1973, SI Units and Recommendations for the Use of Their Multiples and of Certain Other Units,* ANSI, 1973.

THE RELATION OF FORCE TO MASS IN THE SI SYSTEM

Ivan W. Roark

ABSTRACT

The dilemma in the force (weight) mass system is created by the engineers' choice to designate units of force in terms of weight (pounds) which is the gravitational attraction between masses. Weight is the most identifiable property of mass but it varies from place to place on this Earth so that calibration of measuring instruments must be confined to a specific area. The dilemma exists in the minds of those who are unfamiliar with computations involving units of mass, time, length and force. Since there is a constant of proportionality between mass and force they are two distinct dimensions. This paper is written to show the relation of mass to force in the SI system and how the term Newton resolves the semantics dilemma.

MASS

Newton's second law of motion may be stated as follows: If the resultant force acting on a particle is not zero the particle will have an acceleration proportional to the magnitude of the resultant and in the direction of the resultant force.

Salient factors herein presented are particle, force, acceleration, and direction. These are worthy of examination.

A particle is *something:* it occupies space; it has mass; it *is* mass. Anything which occupies space is mass. The worthy attributes of mass are: 1) that it has density, the concentration of the substance of mass per unit volume; 2) there is a mutual attraction between masses which is proportional to the product

---

Ivan W. Roark Coordinator of Technology, Tulsa Junior College, Tulsa, Oklahoma.

of the masses divided by the square of the distance between the masses; it has 3) acceleration which is proportional to the magnitude of the force and 4) motion which is in the direction of the resultant force. There are other dynamic properties which are beyond the scope of this paper.

FORCE

A force is that which, when applied to mass, causes accelerated motion: if no force is applied there is no acceleration. This applies to mass in static equilibrium or mass moving with constant velocity. If there is no change in velocity there is no force acting according to Newton's first law of motion which may be stated as follows: if the resultant force acting on a particle of mass is zero it will remain at rest (if originally at rest) or it will move with constant speed in a straight line (if originally in motion).

A force is as proportional to mass as the acceleration it imparts to mass when applied. It may be stated by the mathematical formula:

$$F = ma \qquad (1)$$

ACCELERATION

Acceleration may be defined as a change in speed of a particle. Common usage suggests acceleration as an increase in speed and deceleration as a decrease in speed. The dimensional properties of acceleration involve units of length, symbol *m* and time, symbol *s* (seconds) in the following manner:

Velocity (speed) equals metres per second.

$$V = m/s \qquad (2)$$

Distance equals velocity multiplied by time in seconds, where V is the average speed.

$$d = V \cdot s^{*} \qquad (3)$$

Acceleration, a, equals ($V_i - V_o$) per second at the end of the first second if V_i is the instant velocity at the end of the first second and V_o is the initial velocity. Thus, if V_o = zero,

$$a = V_i/s/s \qquad (4)$$

at the end of the first second.

*The dot is the symbol for multiplication.

By substituting for V (equation 2) we have

$$a = m/s/s = \frac{m}{s} \times \frac{1}{s} = m/s^2 \qquad (5)$$

DIMENSIONS

Any force system contains four dimensions: mass, length, and time units may be arbitrarily chosen but the unit of force is fixed as a proportion to the unit of mass. Thus, force is *not* equal to mass.

It can be easily demonstrated that the harder you throw the farther the ball will go. Obviously, since the mass of the ball does not change, the harder you throw the ball the greater will be the acceleration required to increase the distance traveled.

Do not confuse acceleration with either velocity or distance. If a mass is accelerated from rest to a velocity of 10 ft. per sec. in one second the final speed is 10 ft. per second but the rate of acceleration is 10 ft. per sec. per sec. The displacement, however, will be the average velocity for the time duration of one second, thus:

$$V_o + V_{10} = \frac{0 + 10 \text{ Ft. per Sec.}}{2} = 5 \text{ feet.} \qquad (6)$$

ABSOLUTE SYSTEMS

In order to quantify physical phenomena of nature a system of basic unit dimensions must be established. If they are chosen so that there is a (1 : 1) one-to-one proportionality between the units, the system is said to be a coherent system and the calculations are simplified. Otherwise a constant of proportionality must be established before valid calculations can be made.

The original unit dimension, length, was, inadvertently, arbitrarily chosen. The French intended the metre to be one ten-millionth of the distance from the equator to the pole. Their calculations were, alas, in error so the metre is now determined in terms of wavelengths of the gas krypton-86. It is sad that a length could not have been chosen to be more commensurable with the inch.

So we have for the dimension length, metre (m). Area is a derivative of m, i.e., $A=m^2$. Volume, also, is a derivative: $V=m^3$. The unit dimension of mass, kilogram (kg), equals the mass which has the volume of one cubic decimetre (dm^3) of water at its greatest density, 4°C at standard atmospheric pressure. The concentration of mass in water at this temperature and pressure is the unit of density (1). The third unit, the unit of time, was originally a specific period of the time duration of the year 1900 but it is now calibrated in terms of the duration of radiation periods of the atom cesium-133. The fourth unit, the unit of force, newton is defined as the force which gives an acceleration of 1 m/s^2 to a mass of 1 kg—the mass of 1 dm^3 of water. From this we can write

$$1 \text{ newton} = (1 \text{ kg})\ (1 \text{ m/s}^2) \qquad (7)$$

This system can be applied anywhere in space, on Earth, on Mars, on the moon, with exactitude.

GRAVITY SYSTEM

Engineers prefer to work in a system based on the acceleration of gravity as it acts upon a particle of mass in a vacuum, (free fall). Since gravity varies from place to place on the Earth's surface, a specific place, at sea level, latitude 45°N, is selected for calibration of instruments. At this point the acceleration of gravity, g, is equal to 9.80665 metres per second per second.

$$g = 9.80665 \text{ m/s}^2 \qquad (8)$$

If this value for acceleration is substituted in formula (7)

$$F = 1 \text{ kg} \cdot \text{m/s}^2 \qquad (7)$$

we have

$$F = 1 \text{ kg} \cdot 9.80665 \text{ m/s}^2 \qquad (9)$$

But the force we desire is that force which will accelerate 1 kg · 1 m/s^2 instead of 9.80665 m/s^2. This will establish coherency in the system. By dividing equation (9) by 9.80665 we have

$$\frac{F}{9.80665} = \frac{1 \text{ kg} \cdot \cancel{9.80665} \text{ m/s}^2}{\cancel{9.80665}} = 1 \text{ newton}$$

This can be written

$$F \times \frac{1}{9.80665} = 1\ \text{kg} \cdot \text{m/s}^2 = 1\ \text{n} \qquad (10)$$

Thus, the constant of proportionality $\frac{1}{9.80665}$ is the ratio of the accelerative force to mass in the SI system. This is approximately one to ten. This means that the unit of force, newton, is the force that will accelerate 1 kg of mass 1 m/s^2.

KILOGRAM

What is a kilogram? It is defined in ASTM Standard Metric Practice Guide E 380−70:

The kilogram is the unit of mass; it is equal to the international prototype of the kilogram (adopted at the 1st and 3rd CGPM 1889 and 1901).

So now you know what a kilogram is! It is the only tangible unit in the SI system: a prototype. Why? Simply because it is made of stable material which possesses the one quality of mass, (WEIGHT) which is readily identifiable.

WEIGHT

The weight, the gravitational attraction, of the mass of one cubic decimetre, 1 dm^3, of water at 4°C at standard atmospheric pressure is the weight of the prototype, 1 kg. Weight can be equated on balance scales which nullify the accelerative force of gravity. Spring scales are not accurate if moved from place to place but calibration by weights will give sufficient accuracy for simple industrial use.

DILEMMA

The force-mass dilemma exists in both the inch and the metric system because engineers equate both mass and force in units of weight. It is no dilemma to them; they understand the computation involved. To others it is an expression involving semantics and propriety. This double usage of weight for two basic units is clarified in the SI by assigning a name newton to the unit of force, but the practice still continues as a result of the use of kilogram for force in other metric systems. The SI system is rationalized by permitting the use of kilogram-force which implies the use of the conversion constant of proportionality 1 : 9.80665, thus:

$$1 \text{ kgf} = 9.80665 \text{ n (newton)} \qquad (11)$$

Note: there is no plural of *newton.*

In the inch system the gravitational acceleration is 32.17 ft./s², thus:

$$a = 32.17 \text{ ft/s}^2 = 9.80665 \text{ m/s}^2 \qquad (12)$$

How sad that, since the length of the metre, m, was arbitrarily chosen, its constant of proportionality is so incommensurable with the inch. Since length, m, was arbitrarily chosen, why not make the constant one to ten? But it is too late! The conversion factor for industry is one inch equals 25.4 millimetres.

$$1'' = 25.4 \text{ mm} \qquad (13)$$

If we substitute the acceleration of gravity 32.17 ft/s² in formula (1), F = ma we have:

$$F = m \text{ (mass)} \cdot 32.17 \text{ ft/s}^2 \qquad (14)$$

If we equate force with weight, F = W, and choose the unit of acceleration to be one pound, the force that will accelerate a mass one foot per second per second, we have, by substitution in formula (1):

$$F = W = (\text{unit mass in wt.} \times 32.17) \cdot 1 \text{ ft/s}^2$$

This in simplified form is:

$$1\#f = 32.17\# \cdot 1 \text{ ft/s}^2 \qquad (15)$$

This may be read: a one-pound force will accelerate a mass of 32.17 pounds one foot per second per second. Thus, in the inch system, the unit of mass, by weight, is equal for 32.17 pounds. This is known as a slug—a term little used, being better known in other semantic forms as fish bait or a density stabilizer in a socialized activity. It follows that,

$$\frac{W}{32.17} = \text{mass in slugs} \qquad (16)$$

Your weight, in slugs equals your weight in pounds divided by 32.17. Corrected, this is your MASS in slugs.

It is agreed that when kg is converted to pounds,

$$1 \text{ kg} = 2.2\#$$

Since 1 kgf = 9.80665 n (one kilogram-force = 9.80665 newton)

$$9.80665 \text{ n} = 2.2\# \qquad (17)$$

By dividing by 9.80665, equation (17) reduces to

$$1\text{n} = .2243\# \text{ (almost } \tfrac{1}{4} \text{ pound force).} \qquad (18)$$

In conclusion, there is no real dilemma: it is mostly a semantic problem. It is to those who must calculate. Those who discover the advantage of the SI system will determine the system which will prevail. Until that time the computer will convert to many systems, the choice of which will depend upon the convenience of computation. It would be regrettable if it ever assumes a legal aspect.

INFORMATION SOURCES

American Society for Testing and Materials, *ASTM Standard Metric Practice Guide (A Guide to the Use of SI—International System of Units),* Philadelphia, PA, ASTM, 1972.

Barbrow, L. E., "The Mass-Weight Dilemma," Reprinted from *The Science Teacher,* October, 1972.

Beer, Ferdinand P. and Johnson, E. Russell, *Vector Mechanics for Engineers,* (McGraw-Hill, 1962), p. 440.

Wolf, Ernst, *Metrication for Engineers,* Dearborn, Michigan, Society of Manufacturing Engineers, 1974.

HISTORY OF OUR MEASUREMENT SYSTEMS

Richard A. Kruppa

> Stand at the door of a church on Sunday, bid 16 men to stop, tall ones and short ones as they happen to pass out as the service is finished, then make them put their left feet one behind the other and the length thus obtained shall be a right and lawful rod, and the 16th shall be a right and lawful foot.[1]

This 16th century German definition of the rod and foot suggests the arbitrary nature of what we know as our customary system of units. The system is comparable to the old house to which additions were added from time to time—each with its peculiar architectural style and different construction materials. Our customary system of measurement has been modified and added to by cultures dating back to Babylonian and Egyptian civilizations. Our customary system is not greatly different today from that brought to the "New World" by the English colonists.[2]

The metric system is much newer. About 300 years ago a Frenchman devised a measurement system based upon a unit of length defined as one ten-millionth of the distance between the pole and the equator measured on a parallel running through France and Spain. This unit was called the metre.[3] The modern metric system or International System of Units (SI) was established in 1960 by the General Conference of Weights and Measures, a diplomatic organization made-up of 36 nations including the United States.[4] It is the International System of Units (SI) which is being adopted world wide.

Richard A. Kruppa, Ph.D. Associate Professor, Department of Industrial Education and Technology, Bowling Green State University, Bowling Green, Ohio.

In the United States the metric system has been legal for trade and commerce since 1866, and since 1893 our pound and yard have been defined in terms of the metre and kilogram. To date, every major industrial nation except the United States has adopted, or is officially committed to adopting SI. But, without question, the United States is adopting SI at an ever increasing rate though we have no national policy for a systematic, orderly, and efficient adoption plan. One need only watch the various metric-oriented newsletters to sense the momentum of the change taking place.

LEGISLATION AND METRICATION

Between the years 1968 and 1971 the United States Department of Commerce conducted the U.S. Metric Study. It concluded that the nation should move deliberately and carefully over a ten year period to metric conversion to be coordinated, but not paid for, by the Federal Government.[5]

To date, Congress has not acted to allow such a program to take effect. Less than a year ago HR 11035, the House Metric Conversion Bill, was defeated 240 to 153.[6] The *Metric Reporter* however, has suggested that it is likely that we will see passage of a metric bill by the new 94th Congress by summer. The reasons given for the likelihood of passage are:

1. that the objections to the Bill which came before the 93rd Congress by labor and small business are likely to be removed from the new bill.
2. the youthful and "progressive" character of the heavily Democratic Congress.[7]

On the other hand, the Government has acted to support metric education. On August 21, 1974, P.L. 93-380, Amendments to the Elementary and Secondary Education Act of 1965, was signed into law. Authorized was the expenditure of $10 million for each of three fiscal years beginning in 1975. Guidelines for participating in the program, if and when Congress appropriates the funds, are presently being drafted.[8]

METRIC INSTRUCTION IN TEACHER EDUCATION

What are the trends in industrial teacher education with respect to metric adoption? The American Industrial Arts Associations (AIAA), American Vocational Association (AVA), and the Home Economics Education Association (HEEA), as part of the Education and Training Sector of the American National Metric Council (ANMC) surveyed approximately fifty teacher education institutions in each of the three fields. From among the 73% return of survey forms, the following information was gathered.[9]

In response to the first question (Table I)—"What has been done in the institutions to teach the metric system?"—it was learned that about 80% of the industrial arts departments have introduced metric instruction into technical courses. About 61% of the vocational teacher education departments reported that metric instruction was being implemented in technical courses.

Table I

Concern #1

What has been done at the teacher education level to teach metric?

	N = 43/50		N = 36/48		N = 28/48		N = 107/146	
	AIAA		AVA		HEEA		TOTALS	
	N	%	N	%	N	%	N	%
a. Metric information added to methods courses	13	30	10	28	9	32	32	30
b. Metric information added to individual laboratory courses	34	79	22	61	14	50	70	65
c. Assigned a person on the department faculty to be a coordinator of metric conversion within the department	18	42	13	36	2	04	33	30

(American Industrial Arts Association)

Table II

Concern #2

What kind of instructional materials have been developed for teaching and/or using metric?

	AIAA		AVA		HEEA		TOTALS	
	N	%	N	%	N	%	N	%
a. Course of study (a complete course on measurement)	12	28	4	11	0	0	16	15
b. Unit plans (for a segment of existing courses)	18	42	16	44	7	25	41	39
c. Instructional sheets (for teacher guides in the department	21	49	13	36	8	29	42	39
d. Assignment sheets (activities for students)	14	33	11	31	6	21	31	29
e. Teaching aids	33	77	19	53	11	59	63	59
1. wall charts	12	28	8	22	3	11	23	22
2. measuring instruments	11	26	8	22	8	22	23	22
3. transparencies	10	23	8	22	3	11	21	20
4. filmstrips	4	10	2	6	1	4	7	6
5. audio cassettes	3	7	0	0	0	0	3	3
6. instructional handouts	13	30	4	11	6	21	23	22
7. conversion tables	10	23	4	11	3	11	17	16

(American Industrial Arts Association)

Question #2 (Table II) asked what kinds of materials have been developed. About 77% of industrial arts departments and 50% of vocational departments indicated that they have developed materials and about 40% of all departments have developed materials and unit plans for metric instruction.

Question #3 (Table III) dealt with how the departments felt about certain aspects of the metric movement. Almost 90% felt that it is the responsibility of teacher education departments to provide pre- and in-service teacher education in the metric system. Half of the departments felt that conversion to the metric system in teacher education should be a five-year process, while only 22% suggested that ten years would be appropriate. 84% felt that it is the responsibility of the professional societies to provide direction to the teacher education institutions.

Table III

Concern #3

How do teacher education departments feel about certain aspects of the metric movement?

*A - Agree *D - Disagree		AIAA N	AIAA %	AVA N	AVA %	HEEA N	HEEA %	TOTAL N	TOTAL %
a. The responsibility of teaching metric to pre-service teachers is that of teacher education.	*A:	41	96	33	94	18	64	92	87
	*D:	2	4	2	6	9	32	13	13
b. The responsibility of teaching metric to in-service teachers is that of teacher education.	*A:	41	96	31	88	21	75	93	87
	*D:	2	4	2	6	6	31	11	10
c. Additional funding support is needed in order for our department to teach the concepts of metric.	*A:	26	61	25	74	16	57	67	63
	*D:	17	39	9	26	10	36	38	37
d. Additional funding support is needed in order for our department to convert to metric measurement.	*A:	36	86	27	79	15	54	78	76
	*D:	7	14	7	21	11	39	25	24
e. The leadership for introducing metric conversion in teacher education should come from the college of education rather than from the department.	*A:	13	30	9	27	5	18	27	25
	*D:	29	67	24	63	15	54	68	64
f. It will take our department about five years to complete the conversion to metric.	*A:	24	56	18	58	13	46	55	51
	*D:	16	37	13	42	10	36	39	36
g. It will take our department about ten years to complete the conversion to metric.	*A:	11	26	11	33	2	7	24	22
	*D:	30	70	22	67	19	68	71	66
h. National professional associations should take the leadership in providing direction for teaching metric at the teacher education level.	*A:	39	91	27	82	24	86	90	84
	*D:	2	6	6	18	3	11	11	10

(American Industrial Arts Association)

The fourth question (Table IV) asked what services and resources would be helpful. While 87% responded that instructional aids and materials were needed, 75% desired transition guidelines.

Table IV

CONCERN #4

What services or resources would be useful to your department in your efforts to move toward metrication?

	AIAA		AVA		HEEA		TOTAL	
	N	%	N	%	N	%	N	%
a. National task force	15	35	10	28	5	18	30	28
b. More articles published	23	54	12	36	8	29	44	41
c. Packets of aids and materials	40	95	28	78	25	89	93	87
d. Consultant services	13	30	11	31	6	21	30	29
e. Guidelines for transition to metric	33	77	25	69	21	75	79	74
f. Staff seminars	26	60	23	64	12	42	61	57

(American Industrial Arts Association)

The momentum of change in teacher education is clearly established. Instruction in SI is underway in the colleges and universities.

COSTS OF CONVERSION

Few conversion cost studies have been reported in the literature of industrial education. Nelson[10] reported that an analysis of costs has been conducted at Stout State University, and he compared it to cost estimates produced in the U.S. Metric Study. The Metric Study staff estimated the conversion cost to be less than 5% of the original cost of tools and equipment. Their estimates range from about $100 for technical laboratories in electronics (already using SI units) to about $20,000 for a tool and die shop (Table V). Estimates presented by Nelson were higher than those of the U.S. Metric Study, but were established for different numbers of work stations.

Table V

Chart: Estimated Costs for Converting Selected Labs to Metric

Area	Number of Stations	Total Cost ($)
Auto Body	12	400
Auto Mechanics	16	4,400
Carpentry and Cabinet Making	20	1,250
Drafting and Design	20	300
Electronics	20	100
Graphic Arts	15	1,000
Machine Shop	15	18,300
Small Engine Repair	15	500
Tool and Die	15	20,000
Welding	16	3,300

From U.S. Metric Study Interim Report: Education, National Bureau of Standards Special Publication 345–6, July, 1971 ($1.75)

Bychinsky[11] reported a study performed at Ferris State College. Costs of converting 50 machine tools, measuring and gaging tools including precision instruments, and cutting tools were estimated at $60,000 over five years. First year conversion cost estimates included a lathe at $400, a milling machine at $495, and a surface grinder at $275.

Another estimate may be derived by assessing costs of tool kit conversions in the trades. Tyrell of the Canadian Metric Commission speculated ". . . it is now tentatively projected (that) the cost of replacement or supplementation of tool kits may range up to or above $1000 dollars."[12] It appears clear from my vantage point that conversion cost estimating will vary widely reflecting the enthusiasm of the conversion effort.

IMPLEMENTATION AND RESOURCES

Implementation strategies need to be developed at the teacher education level and in the public school shops and laboratories. Much advice is available to those desiring to create change.

The *Center for Metric Education and Studies* has proposed a scheme for implementating metric programs at the industrial teacher education level.[13] Initially, the department should identify one faculty member to assume the leadership role. The individual would become an expert in SI, he would collect information related to the system, to hardware and conversion devices, and he would become acquainted with instructional materials which are available. The person would also attend meetings, such as this conference, in his quest to become thoroughly immersed in the issues surrounding metric adoption. The department should next establish its role in the change to metrics, including the development of timetables and priorities. Training the members of the department in SI would precede development and implementation of metric instruction in its pre-service program. This will assure impact among emerging teachers. Once the pre-service program is established, the institution should examine the contribution it can make to in-service workshops and other special programs.

How and when shall the teacher in the school shop or laboratory begin the implementation process? The question is perplexing, for the longer we have to wait for a Federally coordinated adoption plan, the longer an individual will be required to work with and use both our customary system and SI. Additionally, while it is not generally recommended to require students to learn and use conversions extensively, we cannot overlook the fact that we will have to live for many years with the problem of conversion.

Chalupsky and Crawford of *The Metric Studies Center* have proposed a scheme for implementing SI.[14] In doing so, they suggest three procedural stages:

1. Introduce initial work in metric units using existing equipment and facilities with appropriate translation. This stage could include the re-marking of tools made to customary specifications. For example, a ½″ wood chisel would be marked 13mm. Twist drills would be marked with their corresponding metric sizes.
   Presumably student activities would be conducted in both systems. Coursework might be structured such that they design, draw, and fabricate some items in the customary system, and later apply the same procedure to the construction of an item in SI.

2. The planned modification of existing equipment with minimal recalibration. For example, students could recalibrate and re-draw pressure gage faces.
3. The replacement of existing equipment. The final stage is also the most expensive one. At this time, an informed instructor should not consider purchasing new equipment without some provision for dual-measurement indication or SI measurement units.

Instructional resources for SI are becoming available at an ever increasing rate. Unfortunately, with the proliferation of these materials one often finds inaccuracy in the specification of units, even in the spellings and abbreviations. One must, therefore, be careful when selecting materials. I would suggest that a person who is about to purchase materials first review ASTM—E 380—*The Metric Practice Guide*[15] or "A Metric Where-to-Find-It."[16]

CHAPTER NOTES

1. *Metric Reporter,* October 4, 1974.
2. U.S. Dept. of Commerce, *Brief History of Measurement Systems,* National Bureau of Standards, Special Publication 304 A, 1972, p. 1.
3. International Business Machines Corp., SI Metric: *The International System of Units* (White Plains, N.Y.: I.B.M., n.d.), p. 6.
4. E. Wolfe, *Metrication for Engineers* (Dearborn, Michigan: Society of Manufacturing Engineers, 1974), p. 6.
5. U.S. Dept. of Commerce, *Technical News Bulletin,* National Bureau of Standards, Reprint (September 1971): 2.
6. L. Sokol, ed., "Metric Legislation," *Metric Association Newsletter,* (May 1974):1.
7. "Metric Legislation and the 94th Congress," *Metric Reporter* (December 13, 1974):2.
8. "U.S.O.E. Developing Criteria for Metric Education Grants," *Metric Reporter* (December 27, 1974):1.
9. L. Kabakjian, "A Look at Metric Education in Teacher Preparation," *Man/Society/Technology* 34 (1974):19.
10. O. Nelson, "Is Metric a Measure of Pain?" *Industrial Arts and Vocational Education* 61 (1972):23.
11. S. Bychinsky, "Metricorner: Conversion Price Tag," *School Shop* 34 (1974):12.
12. "The Canadian Labor View of Metric Conversion," *Metric Reporter,* (January 10, 1975):4.
13. J. Lindbeck, "Going Metric in Teacher Education," *Man/Society/Technology* 34 (1974):9.
14. A. Chalupsky and J. Crawford, "The Cutting Edge of Metrication," *School Shop* 33 (1974):50.

15. American Society for Testing and Materials, *Standard Metric Practice Guide* E 380-72 (Philadelphia: ASTM, 1972).

16. R. A. Kruppa, "A Metric Where-to-Find-It," *School Shop* 34 (May, 1975): 36–37.

METRICATION AND INDUSTRIAL EDUCATION LABORATORY EQUIPMENT

Armand M. Seguin

That the metric system of measurement is coming to the United States is a moot question and will not be discussed here. The existence of this Second International Conference on Metric Education and others like it certainly attests to the changing system of measuring in our country. Also, the existence of a session like this, devoted to industrial education, attests to the involvement of industrial education in metrication. The metrication process can be divided into two sections: one involving software and the other hardware. Software can be referred to as books, kits, games, and the like and there seem to be kilograms of this material available. Indeed, the exhibitors' wares at this convention are a good example of the quantity and variety of software on the market. However, we in industrial education are only peripherally interested in most of the software items because our instruction centers on applications in the real world. I will be discussing the hardware aspect of metrication as it pertains to industrial education. This aspect of metrication is much more expensive and complicated than that dealing with software.

It might be apropos to digress for one minute and consider who we in industrial education serve. According to *Vocational-Technical Terminology*[1] the term "industrial education" includes the fields of industrial arts, vocational-industrial and industrial-technical education. The field of industrial arts is part of the general education program which aims to prepare people for life. The future will undoubtedly include metric measuring far more than it does today and thus we can easily justi-

---

Armand M. Seguin, Ed.D. Assistant Professor, School of Industrial & Technical Studies, Jackson State University.

fy teaching the metric system. The argument for vocational-industrial and technical education must be somewhat different as these students are preparing for contemporary industrial employment. My own estimate is that only about 5% of these workers now use part of the metric system in their present work. However, since modern industry is rapidly converting it appears inevitable that many of these people will be using the system in the foreseeable future. Thus, it is incumbent upon all of us in industrial education to instruct our students in using the metric system of measurement.

In gathering the information on metric laboratory equipment for this paper, I sampled several different sources. My efforts were not quantified, nor did I attempt to survey all possible sources. The four methods that I used to gather the information are as follows:

1. Letters to manufacturers of laboratory equipment (approximately 50 firms);
2. Letters to major supply houses (like Broadhead-Garrett);
3. Contacts with local industrial supply houses; and
4. Personally examining catalogs containing laboratory tools and equipment.

The overall response from manufacturers was mixed. Some firms are apparently not aware that metrication will affect them, and others are seemingly ahead of certain markets by offering items that few people are buying. The major supply houses seem well aware of metrication but thus far market few items outside of software and measuring tools. The local supply houses were aware of few metric sizes though they often sold "foreign" bolts, tires and whatever. A careful examination of supply catalogs will reveal many metric-sized items. However, the total picture easily shows that the amount of software available to a mathematics educator far overshadows the hardware available to the industrial educator.

The typical industry conversion plan covers a ten year period and involves three separate phases. Lowell Foster[2] described Honeywell's program as including Alert, Adaptation and Applicability. As shown in Figure 1, the first phase lasts two years and involves planning, designing, and making specifications. The Adaptation phase typically lasts three years and involves

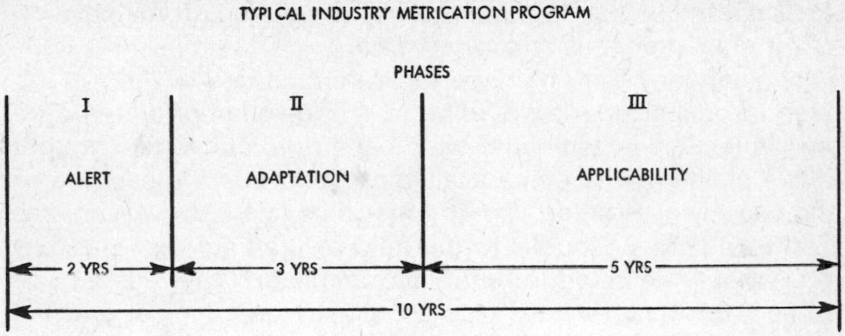

Figure 1

actual production of new products in metric standards. During this phase up to 50% of a firm's products may be metric. The third and last phase lasts five years and involves converting the remaining products to metric standards. In some cases certain products may never change but will merely be phased out when they become obsolete. The plan has obvious implications

Figure 2

READING A METRIC MICROMETER

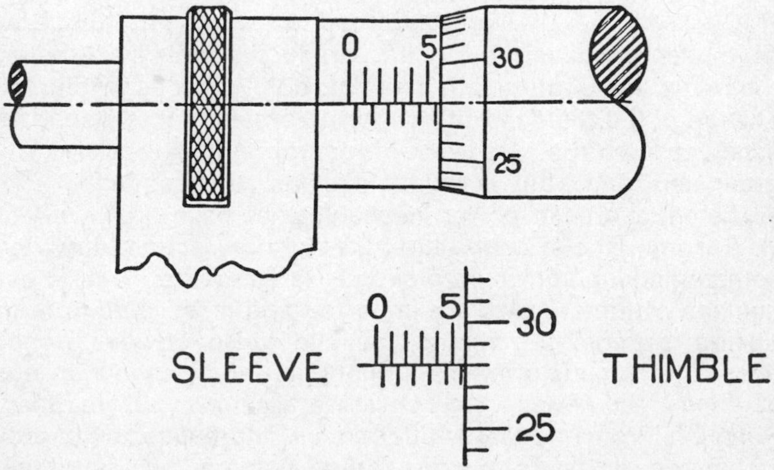

READING 5.78mm

for industrial education, particularly since many firms have already embarked on similar programs.

In nearly any firm, the area of drafting is one of the first affected and just as one would hope, a large number of items are available. Scales, templates, grid paper and converters are but a few of the metric items available. Firms offering such items include Alvin, Bruning, Central Instrument, Hunter Associates and Post. The textbooks in the field of drafting were the only ones that I attempted to randomly sample and they seem to be in need of vast revision. Figure 2 illustrates a typical drawing that uses identification letter dimensioning and shows both metric and English dimensions. It is well to remember that not only will dimensions change but also the method of projection, certain of the dimensioning methods and tolerances. With metric dimensions and standards it is likely that drawings may truly become the "Universal Language" that drawing teachers have so long claimed.

The whole area of measuring tools is also one of the first areas affected. Again, manufacturers of micrometers, vernier calipers, tape measures, and similar items are already in production. Firms like Lufkin, Mitutoyo, Stanley, and Starrett have nearly any measuring device needed. Figure 3 illustrates the ease of reading a metric micrometer, particularly to anyone who can read an inch scale, and underscores the logic of converting to an all-decimal system of measurement. In a letter, F. H. Clarkson, Jr.[3] of Starrett Company states that, "Since we have been manufacturing metric graduated tools since before the turn of the Century, our materials don't reflect a revoluntary change in the direction of the metric system." It is also interesting to note that Stanley Corporation already has a metric graduated square but as yet markets no framing square.

The entire area of power mechanics has been greatly affected by the metric system. About 25% of our cars are imported, some American engines are already rated in litres, a high percentage of motorcycles are imported and most large farm-industrial tractors are metric. Despite these obvious trends, about the only metric items available are tools used with nuts and bolts. The news is good however, as, among others, Sears, Snap-On, Proto and Wards offer a nearly complete line of tools for metric fasteners. Among other readings, pressure and power ratings will change but few equipment manufacturers reflect the trends. A relatively small firm, Megatech Corpora-

MASTER DIMENSIONING

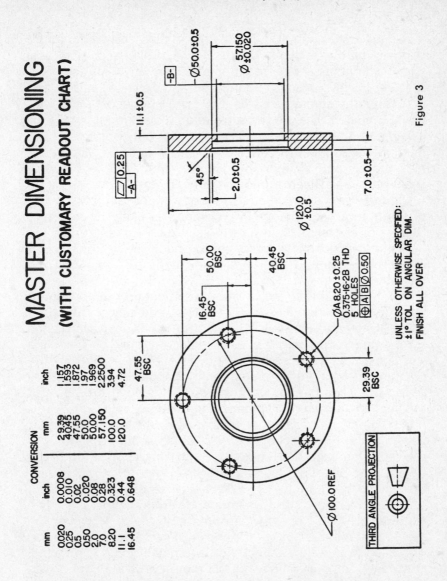

(WITH CUSTOMARY READOUT CHART)

CONVERSION

mm	inch	mm	inch
0.020	0.0008	29.39	1.157
0.25	0.010	40.45	1.593
0.5	0.02	47.55	1.872
0.50	0.020	50.0	1.97
2.0	0.08	50.00	1.969
7.0	0.28	57.150	2.2500
8.20	0.323	100.0	3.94
11.1	0.44	120.0	4.72
16.45	0.648		

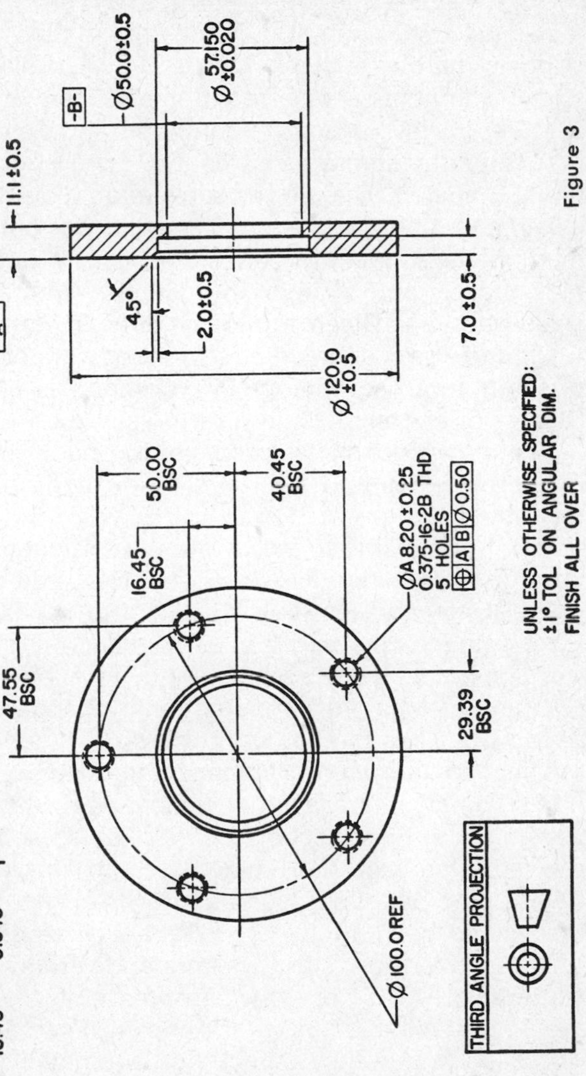

Figure 3

tion, markets a metric dynamometer, but the well-known Sun and Marquette Corporations offer nothing metric that I could find. Along these general lines, the Ken Cook Company sells an automated teaching system that uses a Kawasaki motorcycle. Since the Kawasaki is metric, the training includes lessons in the metric system as it applies to the motorcycle, and it uses the "cold turkey" approach.

The machine shop area is one that is blessed with an abundance of hardware items. Taps and dies, drill bits, cutting tools and similar items are offered. Some of the many manufacturers are Greenfield, Hanson, Regal-Beloit and Zelenda. Lathes and other equipment may be purchased from Bridgeport, Cincinnati and Clausing. The area of conversion dials and kits made for existing equipment is also active with not only original equipment manufacturers but firms specializing in such, like Sipco-Mic, Stavely and Vickers. In addition, Wilton markets metric-sized vises, and Diacro-Houdaille offers metal shears.

I found the manufacturers of woodworking equipment surprisingly aware of metrication. Companies like Rockwell and Stanley offer complete lines of tools and equipment. Since the area of construction will probably be one of the last affected, it is interesting to find a relatively high degree of awareness in the industry.

The graphic arts area is aware of metrication but appears to be proceeding cautiously. Firms like Addressograph-Multigraph, Ditto and Kodak market metric items. However, since most American printing is in English and this severely limits the world market, there seems to be no rush to supply metric equipment. Most ISO (International Standards Organization) paper sizes can already be printed on American presses and the final copies can be trimmed to any size.

In summary, there is metric hardware available in America for use in industrial education laboratories. In some instances, the education market has lagged behind the suppliers. Chuck Fehr[4] of the Satterlee Co. says that "Requests for metric equipment have been spotty. . . . We anticipated good movement on dual dimensional micrometers and vernier calipers . . . two years ago. Our sales were disappointing. . . ." However, on the whole, manufacturers are aware of metrication and I can only assume that they expect their equipment to be in demand. Obviously, some areas have much more hardware than others but it is generally available and more will be ready when demand increases.

From the information learned in making this presentation and my experiences in metrication, my recommendations for industrial educators are as follows:

1. Think metric;
2. Teach metric. Don't wait for textbook changes;

3. Teach both systems, separately, and convert (with converters) *only* when necessary;
4. Purchase metric (not dual) measuring tools;
5. Emphasize metric in drafting;
6. Purchase graduation plates and kits for large equipment;
7. Develop metric projects; and
8. Cultivate a "feel" for a centimetre, kilogram and litre by using metric hardware and standards in the industrial laboratory.

CHAPTER NOTES

1. *Vocational-Technical Terminology.* Washington: American Vocational Association, 1971.
2. Lowell W. Foster, "A Look at Honeywell's Approach to Metrication," *Metric News* (March/April 1974): 34–35.
3. F. H. Clarkson, Jr. Letter to Dr. Armand M. Seguin dated January 16, 1975.
4. Chuck Fehr. Letter to Dr. Armand M. Seguin dated January 24, 1975.

WHY TOTAL IMMERSION?

Willard F. Reese

My education as a science teacher has caused me to be conversant with metric units of weights and measurement for many years. (It is true, like many of you, I've had to do some relearning in order to become SI-Metric oriented.) Because of my training, while living in or visiting metric countries I had little difficulty accepting laboratory weights and measures in the market place. In fact, I was delighted to do so. Not everyone is fortunate enough to have this type of background. Coming from a bilingual country as I do I may equate my language handicap (I can only understand the print on one side of my breakfast cereal box) with the difficulties encountered by an individual who only understands one system of weights and measures in a society where he encounters two.

It is not my place nor the business of this Conference to discuss the pros and cons of bilingualism but, keeping to the analogy, an important lesson may be drawn from this experience. Children in Western Canada are taught French from the primary school throughout their elementary grades and many go on to take three years of French in high school. Traveling with young people thus "educated" was very revealing. Most of them had difficulty communicating on even a very elementary level. The one young lady who was able to act as our translator was from Dallas, Texas, and had never had a French lesson, per se, in her life! However, she had been on an exchange with a French family in a little village where French was the only language spoken!

I had the good fortune of spending a sabbatical leave, during the 1972–73 academic year, in the United Kingdom. English

---

Willard F. Reese, Ed.D. Professor of Science Education, University of Alberta, Edmonton, Canada.

educators and metric consultants concurred with their North American counterparts in that they believed that total immersion is the best technique for learning the SI-Metric system. This "cold turkey" approach seems eminently reasonable to us, but what research evidence do we have to support it? My British friends fell back on another analogy (i.e., when Sweden changed from left side to right side of the road, instead of more accidents there were far fewer than usual for the first 24 hours) which was not very appropriate or convincing in terms of metrication.

Nowhere in the literature could I find research evidence to support the total immersion hypothesis. Last spring while preparing a metric workshop I decided to try a simple little experiment.

The forty teachers who were to participate in this workshop were all pretty much on square one as far as knowing the metric system was concerned. As they entered the workshop room, every other teacher was asked to step on the bathroom scales to the right and read the weight (mass) in metric units. The next teacher was asked to go left and do the same thing. Both scales were identical except that the one on the right had only kilogram graduations, whereas the one on the left showed both pounds and kilograms.

An hour into the workshop, the teachers were asked "How many of you can honestly remember your weight in kilograms?" 16 out of 20 who weighed on the kilogram-only scale remembered, but only 9 out of 20 who weighed on the double scale still knew their weight in kilograms.

The results of this little pilot study were so encouraging that I decided to experiment with a large class of students in their first year in the Education Faculty at the University of Alberta. Seventy-nine students participated in this study. As they entered the large lecture hall, the first was sent to the right (blue side), the second to the left (red side) and so on. The two sides were identified by large pieces of colored cardboard. In addition to weighing on the bathroom scales, the students were asked to find their height in metric units. On the side with the dual-dial bathroom scales, I had placed two metre sticks one on top of the other against the wall. On the side of the metric only bathroom scales, I had two more metre sticks plus two yard sticks next to them and similarly arranged.

No explanation was offered. The students were asked to find

their metric height and weight. When the students had weighed in and checked their height, they sat down and listened to an hour long panel discussion on educational administration.

Following this, the students were provided with paper and asked two main questions. The questions with the number answering each one were as follows:

Questions	Red (N=37) Yes	Red (N=37) No	Blue (N=42) Yes	Blue (N=42) No
1. Do you honestly remember your weight in kilograms?	16 (43%)	21 (57%)	35 (83%)	7 (17%)
2. Do you honestly remember your height in centimetres?	35 (95%)	2 (5%)	19 (45%)	23 (55%)

After each question, students were also asked to tell whether or not they knew their metric weight and height prior to today. In the red group, only two said that they had known their weight prior to the experiment and only one person in the blue group knew it in advance. One person said he knew his height in centimetres in advance. He was in the red group. These few individuals do not appreciably affect the proportions indicated above.

If we apply the statistical test of a difference between two independent proportions to the responses to question one, we find that the difference is highly significant ($Z = 3.7$; $p < .001$). The proportion of the subjects who could recall their weight (mass) in kilograms was significantly greater in the group who weighed on the metric-only scales (83%) than in the group who weighed on the dual-dial scales (43%).

Similarly, the proportion of students who could recall their height in centimetres was significantly greater in the group that measured themselves only in centimetres (95%) than in the group that took their height in both British and metric units (45%) ($Z = 4.7$; $p < .001$).

While this experiment was not as well controlled as it might have been, the results are so one-sided as to suggest that the cold turkey hypothesis is valid. The most efficient way for us to help our students to "think metric" is to immerse them in the metric system and avoid, as far as possible, any reference to the British system of measurement. This little experiment indicates that when people are faced with both systems they concentrate on the familiar one and ignore the other. Conversely, when forced to measure themselves in the metric system only, they are forced to think in terms of metric units!

It's not often that a science educator gets an opportunity to try out his poetry on a capitive audience so let me leave you with just one verse of my poem entitled "Metre Meter":

Renounce the ounce,
  Set the hounds
On the pounds.
  They're useless like tinker's dams
For the world is using kilograms!

IMPLEMENTING METRIC MEASUREMENT

Charles Eicher

INTRODUCTION

The inevitability of the movement toward adoption of the metric system of measurement as a primary system by the United States is considered in the first part of this paper. Facilitating this transition from the English system of measurement to the metric system in the least disruptive manner possible will require the collaborative efforts of various institutions and agencies. Educational institutions share a major portion of the responsibility for this transition in helping their clients acquire competencies which will enable them to communicate measurement with the metric system as their primary language of measurement.

The second part of the paper suggests possible metric learning experiences throughout the elementary and secondary levels of school in considering the metric system as a curricular topic which should be dealt with broadly throughout the total school program rather than as a topic reserved for the secondary science classroom. At this point there is no attempt made to describe a complete model program at either level, but to suggest a broad program based upon an inductive approach. A program of this nature will involve many teachers who have had little or no contact with the metric system of measurement. In-service activities providing these teachers with an opportunity to learn the metric system of measurement inductively are considered essential if those same teachers are to provide their clients with similar programs. Several activity-oriented learning centers designed for in-service teacher education workshops on metric measurement are described within the final part of

---

Charles Eicher, Ed.D. Associate Professor of Elementary Education, University of South Dakota.

the paper. Metric measuring tasks for five activity centers are identified in the appendix. The activity centers identified have been used by the writer for several in-service teacher education workshops within the University of South Dakota service region.

A PERSPECTIVE

That the United States is moving toward adoption of the international metric system of measurement is exceedingly clear in view of the evidence about us. News media have been reporting with increasing frequency legislative activity within Congress aimed at speeding this nation's conversion to meters, liters, and kilograms. Signs showing distances in kilometers are already appearing on the highways of our nation. The National Park Service has announced plans to add metric measurements to park signs and brochures. Statewide efforts in increasing the metric measurement activities within the curricula are underway in both Maryland and California. With rapidly increasing frequency the people of this nation will encounter expanding application of the language of metric measurement within their daily lives.

A study conducted by the National Bureau of Standards reported to Congress in 1971 the need for the United States to officially promote the adoption of the international metric system of measurement as its primary system of measurement. It was advised then that conversion from the English system to the metric system be accomplished over a ten-year period in a carefully planned, systematic manner. A bill was introduced in Congress calling for voluntary conversion to the metric system, with subcommittees from both the Senate and the House working on the bill. A Senate bill was passed in 1972 but the House has yet to pass a metric conversion bill. Bills considered have called for a presidentially-appointed board to coordinate industrial and consumer efforts in a voluntary conversion to the metric system over a ten to twelve year period. In bills considered thus far this board does not function as an enforcement agency. It is generally felt that Congress will pass a form of this voluntary metric-conversion bill in the near future. Congress this year did pass legislation authorizing the expenditure of $10,000,000 for each of the fiscal years ending prior to July 1,

1978 for the purpose of facilitating systems for teaching metric measurement to the children of the United States.

Every major nation in the world today has adopted, or is in the process of adopting, the metric system as its official language of measurement. The United States, as an industrial power in a world of increasing trade competition, cannot afford to risk competitive trade advantage by failing to communicate measurements in the international language of the metric system. In many sectors of business in this nation the conversion to metrics is already underway voluntarily as business deals with the problem of competing in a world market. The expense of failing to convert in terms of trade lost is many times greater than the expense involved in changing to the metric system.

Industries providing munitions, weapons and other material to the armed forces of NATO have for several years been operating in terms of international metric units, as have the Armed Forces of the United States in complying with a very real military need of operating defense systems with allies whose weapons and material must be readily interchangeable. The advantages gained by simplifying logistics in this manner are obvious.

The metric system is simpler and more logical than the English system and is more precise as a tool of measurement. Operations within business and industry, which are becoming increasingly more dependent upon a computerized technology, are certain facilitated more quickly and more accurately by the decimal system of metric measurement. Metric measurement is becoming an international system for these reasons and will continue to gain increasingly greater support within a technological society such as ours with its heavy investments in world trade. The move to the metric system is inevitable. The question, then, becomes one of how to facilitate transition from the English system to the metric system with the least amount of disruption and confusion, helping all Americans to achieve functional literacy with this language of metric measurement.

Barriers to full implementation of the Metric System will and do exist even in the earliest of planning stages. It would appear that economic and attitudinal barriers will be the most prevalent and the most difficult to overcome. Pressures to force implementation, concurrently accomplished with an effective educational program, must come from the people and result in a collaborative endeavor by governmental, educational, civic, and occupational institutions at all levels. Although the focus of this paper is upon implementation of the metric system of measurement in terms of educational programs within the schools of our nation, the necessity for a total collaborative effort is viewed as essential for a smooth transition to metrics. The remainder of the paper will suggest points for consideration in teaching metric measurement at the elementary and secondary school levels and will describe metric-measurement activities which have been used at several in-service workshops for teachers.

TEACHING METRIC MEASUREMENT

Children now starting school will graduate into a world of metric measurements. Schools must assume responsibility *now* for programs that will help children achieve functional literacy in the language of metric measurement. Functional literacy should enable a person to communicate ideas relative to metric measures. Competencies for communicating in the metric system should enable one to determine whether or not a statement involving a metric measure is absurd. Thus the statement that the weight of the average adult American male is two hundred kilograms is identified immediately as absurd by the person who thinks metric. Likewise, the statement that my uncle is two decimeters tall is quickly declared preposterous, but the statement claiming a particular automobile's gasoline tank capacity as sixty-five liters is viewed as quite probable. The person who is functionally literate in the metric language does not need to revert to an English-metric conversion table to attach meaning to metric units about which he has read. If he reads or hears that a tower to be constructed is to be twenty meters high, he can quickly visualize the approximate height and it has meaning for him.

Elementary Level

Programs in the elementary school to help children "think metric" should be based upon an inductive learning approach. Numerous activity sessions should involve the children in measuring with the metric system thus enabling the children to generalize concepts, such as the size of the liter. "Four of us together can easily drink a liter of milk during lunch, but I can't drink that much milk by myself." The program should involve metric measuring, metric measuring, and more metric measuring in meaningful, applied situations. There should be no conversion between the English and metric systems. The standing long jump, measuring each child's jump with a meter stick, provides an excellent way for children to develop a functional understanding of meter and decimeter. Marking the face of a bathroom scale with kilograms and having children guess weights in kilograms is an excellent activity.

There are numerous English-system measuring activities found in current textbooks which are quite adaptable simply by measuring in metric units rather than English units. Minimum materials required for a good activity program include centimeter rulers of 20 or 30 centimeters length, numerous meter sticks (marked with decimeters, centimeters, and millimeters), balance scales (with various weights marked in grams) bathroom scales marked in kilograms, and several containers with liter and milliliter markings.

Secondary Level

The metric system has long been an integral part of the science program at the secondary level. Here programs can be given meaning by involving students in measurement activities that are realistically associated with a particular science activity. Measurement then is seen as a function or necessary operation within a broader science project or activity.

There is a very real need to extend the metric system to those areas of the secondary curriculum which have been involved primarily with the English system of measurement. Areas such as home economics, wood shop, machine shop, consumer studies and driver education provide an excellent opportunity for students to deal with the world of metrics.

Teacher Training

Now is the time to begin measurement activities in the metric system. Although science teachers at the secondary level can effectively work with their students in the metric system, most teachers do not feel quite so confident. If *now* is the time to begin, then no further time should be lost in providing pre-service teachers and in-service teachers with competencies to enable them to work effectively within the metric system. An inductive approach should be used with teachers, as they should use an inductive approach with their children, which means measure, measure, measure. Learn to think metric. Use materials for measuring meters, liters and grams. Build a liter container from construction paper. Measure, measure, measure in the *metric* system. An effective person to help provide in-service activities for the school faculty is that science teacher who already thinks metric.

The writer has conducted several in-service teacher education workshops on metric measurement within the University of South Dakota service region. The nature of these workshops has been basically inductive, with teachers involved in metric measuring tasks identified at each of five learning centers. These learning centers, or activity centers, along with their respective metric measuring tasks, are included in the appendix of this paper. The workshops have varied in length, with at least a full day required for the greatest effectiveness.

At the beginning of each workshop there are two short lecture/participation sessions. One centers upon justifications for the transition from the English system of measurement to the metric system. The other is an explanation of the metric system in brief, discussing the basic units of meter, liter and gram, utilizing a hands-on approach in which participants handle several metric objects. Among these are the meter stick, a liter container, a kilogram weight and a gram weight. Prefixes are viewed with respect to their physical effect on the basic metric unit being considered. As an example, if the meter is the basic unit being considered, then workshop participants will physically display the millimeter, centimeter, decimeter, meter and dekameter respectively as the diameter of paper clip wire, the length of a white Cuisenaire rod, the length of an orange Cuisenaire rod, the length of a meter stick and the length of a dekameter strand of rope. Again, this represents a hands-on approach to facilitate greater understanding of the prefix system of metric measurement and to provide a foundational experience for explaining more meaningfully the nature of the powers-of-ten relationship of the prefix system. Through experiences such as these, the logic of the metric system seems to be more readily grasped by those who have admitted very little prior understanding of metric measurement as a logical system.

The greater amount of workshop time involves the participants in measuring tasks at the activity centers. Each of the five activity centers is set up at a different table, at which a participant receives an activity center task sheet to be completed by measuring with metric tools. Upon completion of the task sheets, participants are given answer sheets to check the correctness of their responses. They then move on to the next activity center. This procedure is followed until the workshop participants have been involved in measuring at each activity center. The emphasis is upon measuring with metric tools and not upon converting from one system to another.

Participants by this time have been involved in a great many metric measuring activities and are generally ready to evaluate realistically metric teaching aids with respect to their suitability for an inductive approach at a particular grade level. This final session on evaluation of metric teaching materials also serves as an excellent capstone session for tying loose ends and re-emphasizing the need for an inductive teaching approach.

APPENDIX: ACTIVITY CENTER METRIC MEASURING TASKS

ACTIVITY CENTER I

Metric Linear Measurement

Task Sheet

Complete the following tasks. Do not move to the next activity center until after you have received the answer sheet for these tasks.

1. Compare the meter stick with your yard stick. Which is longer, a yard or a meter?
2. Examine the meter stick, noting that the meter is subdivided into 10 decimeters, 100 centimeters and 1000 millimeters. How many centimeters equal one decimeter?
3. Measure each of the following items and record the measure obtained to the nearest indicated unit.
   a. Length of large paper clip. ________mm (millimeters)
   b. Length of small paper clip. ________mm
   c. Diameter of penny. ________mm
   d. Width of narrow white film leader. ________mm
   e. Width of film on grey reel. ________mm
   f. Width of film strip in green container. ________mm
   g. Length of orange Cuisenaire rod. ________cm (centimeters)
   h. Length of black Cuisenaire rod. ________cm
   i. Length of domino. ________mm
   j. Length of orange rectangular cardboard. ________dm (decimeters)
   k. Width of orange rectangular cardboard. ________mm
4. What is the length of the domino in centimeters (carried out to the nearest millimeter)? ________cm
5. Measure the tallest person and the shortest person in your group (carried out to the nearest centimeter).
   a. tallest person________m (meters)
   b. shortest person ________m
   c. tallest person ________dm
   d. shortest person ________dm
   e. tallest person ________cm
   f. shortest person ________cm
   g. tallest person ________dam
   h. shortest person________dam (dekameters)

6. Measure the width of your thumbprint. __________cm

7. Measure the span of your outstretched hand, from thumb to little finger. _____cm

ACTIVITY CENTER I

Metric Linear Measurement

Answer Sheet

1. A meter is slightly longer than a yard.

2. 10 centimeters equal one decimeter.

3. a. 52 mm

   b. 32 mm

   c. 19 mm

   d. 8 mm

   e. 16 mm

   f. 35 mm

   g. 10 cm

   h. 7 cm

   i. 44 mm

   j. 3 dm

   k. 110 mm

4. 4.4 cm

5. Answers will vary (1.55 - 1.93 m)

6. Answers will vary (2 - 3 cm)

7. Answers will vary (18 - 25 cm)

ACTIVITY CENTER II

Metric Measurement - Capacity

Task Sheet

Complete the following tasks. Do not move to the next activity center until after you have received the answer sheet for these tasks.

1. Place square region A on the graph paper. How many square centimeters (cm^2) are equivalent to area A? How many square decimeters (dm^2) are equivalent to area A? What is the ratio of 1 dm^2 to 1 cm^2? What is the ratio of 1 dm to 1 cm? What happens to the metric powers-of-ten relationship when the measure changes from one dimensional to two dimensional?

2. Compare the size of square region A with the size of each of the five square regions identified as B. Are each of the regions B smaller than, larger than, or approximately the same size as region A? Each region B is how many decimeters wide? ________ How many decimeters long? ________

3. With region B_3 as bottom, fold regions B_1, B_2, B_4, and B_5 up temporarily to form a container. What is the length of this container? __________cm What is the width of this container? ___________cm What is the height (depth) of this container? ___________cm
4. What is the length, width and height of block C in centimeters? length = _________cm, width = ___________cm height = __________cm; Block C is one cubic centimeter (1 cc or 1 cm^3)
5. Compare block C with block D. What is the measure of Block D in inches? Is one cubic inch larger than one cubic centimeter?
6. If container B is folded together, how many blocks the size of C does its capacity equal? How many cubic centimeters does the capacity of container B equal? _________cm^3 How many cubic decimeters does the capacity of container B equal? ___________dm^3 What happens to the metric powers-of-ten relationship when the measure changes from one dimensional to three dimensional?
7. Container B, if fastened together, is roughly equivalent to one liter. Measure one liter of water into the white plastic container and mark the level to which the liquid rises with a felt tip pen. Measure one quart of water into the white plastic container and mark the level to which the liquid rises with a felt tip pen. Is one quart a larger capacity or a smaller capacity than one liter?
8. One cubic centimeter (1 cc or 1 cm^3) is equal to how many milliliters?

ACTIVITY CENTER II

Metric Measurement - Capacity

Answer Sheet

1. 100 square centimeters (100 cm^2)
   1 square decimeter
   Ratio of 1 dm^2 to 1 cm^2 is <u>100 to 1</u>
   Ratio of 1 dm to 1 cm is <u>10 to 1</u>
2. approximately the same size
   1 decimeter wide
   1 decimeter long
3. 10 centimeters long
   10 centimeters wide
   10 centimeters deep
4. 1 cm long
   1 cm wide
   1 cm high
5. Block D = 1 cu. inch, yes
6. 1000 blocks
   1000 cm^3 or 1000cc
   1 dm^3
7. One quart is slightly smaller than one liter.
8. 1 cm^3 = 1 milliliter (1 ml)

ACTIVITY CENTER III (DETERMINING MASS)

Task Sheet

Station # 1 -

Estimate first and then determine the actual weight of each item listed below to the nearest gram.

est. actual

(a) Arithmetic book ________ / ________

(b) Hunt's tomato puree ________ / ________

(c) Hunt's whole tomatoes ________ / ________

(d) Del Monte sweet peas ________ / ________

(e) 1 liter of water (excluding container) ________ / ________

ACTIVITY CENTER III (DETERMINING MASS)

Task Sheet

Station # 1 -

(a) Arithmetic book - 732-733 grams

(b) Hunt's tomato puree - 360-361 grams

(c) Hunt's whole tomatoes - 926-927 grams

(d) Del Monte sweet peas - 575 grams

(e) 1 liter of water = 1 kg (approx.)

ACTIVITY CENTER IV (DETERMINING MASS)

Task Sheet

Station #2

Estimate first and then determine the actual weight of each item listed below to the nearest gram.

est. actual

(a) Nail clipper ________ / ________

(b) Chalk ________ / ________

(c) Ink Pad ________ / ________

(d) Empty can ________ / ________

(e) Plastic base "A" ________ / ________

(f) Electric plug "B" ________ / ________

(g) Piece of lead ________ / ________

(h) Eraser ________ / ________

Station # 3 -

Estimate first and then determine the actual weight of each item listed below to the nearest gram.

	est.	actual
(a) Five cent piece	/	
(b) Sheet of paper	/	
(c) Large washer	/	
(d) Small washer	/	
(e) Small paper clip	/	
(f) Large paper clip	/	
(g) Dime	/	

ACTIVITY CENTER IV (DETERMINING MASS)

Answer Sheet

Station # 2 -

(a) Nail clipper - 15.5 grams

(b) Chalk - 8 grams

(c) Ink Pad - 39 grams

(d) Empty can - 68 grams

(e) Plastic base "A" - 25.3 grams

(f) Electric plug "B" - 26.7 grams

(g) Piece of lead - 22.5 grams

(h) Eraser - 47 grams

Station # 3 -

(a) Five cent piece - 5 grams

(b) Sheet of paper - 4.5 grams

(c) Large washer - 16.5 grams

(d) Small washer - 6.8 grams

(e) Small paper clip - .7 grams

(f) Large paper clip - 1.7 grams

(g) Dime - 2.2 grams

ACTIVITY CENTER V (PART 1)

Metric Measurement - Temperature Measures

Task Sheet (Complete both parts)

At your work station there are three containers, each containing mixtures of water (liquid or solid and liquid). You will also find a thermometer graduated in Celsius (centigrade) degrees and Fahrenheit degrees. Your task is to <u>first</u> estimate the temperature of the contents of <u>each</u> container and <u>then</u> measure and record the Fahrenheit reading and the Celsius reading of each.

1. Container A ⟶ °F = __________, °C = __________
2. Container B ⟶ °F = __________, °C = __________
3. Container C ⟶ °F = __________, °C = __________
4. Give your temperature estimates of the following:
    a. Water freezes at __________°C.

b. Water boils at __________ °C.

c. Comfortable room temperature is about __________ °C.

d. A cold January day in South Dakota. __________ °C.

e. A hot August day in South Dakota. __________ °C.

5. At what point on the Fahrenheit scale would the reading be the same as the Celsius scale?

Metric Measure - "Thinking Metric" (Part 2)

Task Sheet

Respond to each statement below by determining whether or not the statement is an absurdity. In the blank preceding each statement, write either A or R as you interpret the statement as either absurd or reasonable. If the statement is absurd, can you change the amount given to make it reasonable?

_____ 1. Bobby drank 237 milliliters of milk for lunch.

_____ 2. The interstate highway speed limit is 30 kilometers per hour.

_____ 3. My brother in the third grade weighs 120 kilograms.

_____ 4. The nurse determined the boy's temperature to be 37° Celsius.

_____ 5. The six-pack contained 1.5 liters of soda pop.

_____ 6. The envelopes for my Christmas cards measured, on the average, 4 decimeters by 6 decimeters.

_____ 7. Our math book weighs less than one kilogram.

_____ 8. My pen measures 16 centimeters in length.

_____ 9. A newly minted nickel weighs 586 grams.

_____ 10. We enjoyed swimming in the water which measured 85° Celsius.

_____ 11. Our refrigerator door measures 6 meters in height.

_____ 12. The coffee cup holds just slightly more than 4 liters.

ACTIVITY CENTER V

Answer Sheet

Part 1: Temperature Measures

4. a. 0°C

b. 100°C

c. approx. 22°C

d. approx. -30°C

e. approx. 37°C

5. -40°F = -40°C

Part 2: "Thinking Metric"

1.	R	7.	R
2.	A	8.	R
3.	A	9.	A
4.	R	10.	A
5.	R	11.	A
6.	A	12.	A

IN-SERVICE WORKSHOPS FOR METRICATION—ELEMENTARY AND MIDDLE SCHOOLS

Margaret M. Garr

OBJECTIVE: to set up and conduct metric laboratories to familiarize elementary and middle school teachers with the metric system and to provide ideas and motivation for developing "hands on" activities for the classes that they teach.
Two points of view are included:

A. Working directly with suitable metric units of length, area, volume, capacity, mass and temperature; then making tables converting the unit used into equivalent larger or smaller metric units as specified.

B. Comparison of metric measure with familiar units of English measure on the rationale that adults and older children who have spent a lifetime working in the English system will feel more secure in the use of the metric system if they have a basis for visualizing the "new" units in terms of the familiar.

PROCEDURE: Eleven stations are arranged. The stations are identified as A, A-1, B, C, D, . . . J.
After a brief introduction which includes a chart of metric prefixes, with emphasis on powers of 10 to relate the measures within a given class, the members of the group of approximately 50 teachers letter themselves in sequences, A-J. Each group goes to the station identified by its letter and conducts the experiments outlined on file cards.

Margaret M. Garr, B.S., M.A. Mathmatics Consultant, Jefferson County Public Schools, Louisville, Kentucky.

Station A-1 is an extra station. Participants are instructed to move to an empty station as soon as they complete the experiment at the station at which they have been working. In this way, each group moves to another activity without waiting for another group to finish, as there is one more station than the number of groups.

After a period of lab time (varying from 30 minutes to $2\frac{1}{2}$ hours; from one day to 5 days, depending on circumstances), the teachers return to the large group where a recorder from each group, showing examples when appropriate, tells of the experiences and conclusions developed in the first station in which the group worked.

A brief written evaluation follows, and teachers are given handouts of other suggested activities and bibliographies.

DETAILS OF STATIONS

(Written on file cards at each station; keys are available for checking results at the conclusion of the task.)

A. LENGTH

Equipment: Metre sticks; metre tape; 20-cm rulers.

Procedure:
1. Find the length of the diagonal of this file card.
2. Find the width of the table.
3. Find the height of the door.
4. Find the circumference of the yellow plastic coffee can top.
5. Find the length of the room.
6. Find the width of the room.
7. Find the width of this paper clip.
8. Find the thickness of the paper clip wire.

Use the most appropriate metric measure to find each length. Make a chart, converting each measure you found to the other listed units. Draw a ring around the measure you actually used. (mm, cm, dm, m, km)

(NOTE: These activities seem important to help the child recognize the need for various different measures and the simple relationships of powers of ten that exist among them. It also has been observed that many children view width, height, depth, circumference, etc., as *differing* from length; hence the variety in this activity.)

A-1: LENGTH (comparison with English units)

Equipment: Metre stick; yard stick; metre tape; yard tape.

Procedure:

1. Compare the metre stick with the yard stick. Make a simple statement about their relative lengths.
2. Compare the centimetre with the inch. Make a simple statement about their relative lengths.
3. How many centimetres are in one metre? What is an easy way to remember this relationship?
4. How many millimetres are in each centimetre? How many millimetres are in each metre?
5. How many decimetres are in each metre? How many centimetres are in each decimetre? How many millimetres are in each decimetre?

Optional: Draw some conclusions about the numbers of each of these units in a kilometre, which is 1000 metres in length.

6. What statement can you make regarding relationships within metric units of length compared to relationships within English units of length (mile, furlong, rod, yard, inch, $\frac{1}{2}$ inch, $\frac{1}{4}$ inch, $\frac{1}{32}$ inch, etc.)?

B: AREA

Equipment: Metre sticks, 20-cm rulers, square metre of cloth, geoboard and rubber bands, file card with drawing of rectangle 3mm × 7 mm in one corner.

Procedure: Use the cloth which is one square metre in area to help visual area; use the geoboard with rubber bands making rectangular regions of various shapes and sizes to further establish background if desired.

1. Find the area of this file card.
2. Find the area of the table top.
3. Find the area of the floor of this room.
4. Find the area of the rectangular region drawn in the corner on the file card.

*Use what you feel to be the most appropriate metric unit measure for calculating each area. Make a simple table converting each to the other listed metric units. Draw a ring around the measure you actually used. (mm^2, cm^2, dm^2, m^2 . . . km^2)

C. VOLUME

Equipment: Twelve or more metre sticks; masking tape; cubic centimetre blocks (unit cuisenaire rods); cubic decimetre models. (See Appendix I at the end of this chapter.)

Procedure: In a corner of the room, construct a skeleton model of a cubic metre, using the metre sticks and tape.

Place a few cubic centimetre blocks inside the structure; place a few cubic decimetre blocks inside the structure.

1. How many cubic centimetre blocks would it take to fill the cubic metre?
2. How many cubic decimetre blocks would it take to fill the cubic metre?
3. How many cubic metres would it take to fill this room?
4. What part of a cubic kilometre is a cubic metre?

D: LIQUID CAPACITY

Equipment: Water (could use rice, beans, or sand, but liquid helps emphasize the correct idea); 1 standard measuring cup; 1 quart milk container, cut to hold precisely one quart; 1 liter container (1 cubic decimetre) of water-proof cardboard; 1 tray; funnel; towel (in case of spillage). (See Appendix I following this chapter.)

Procedure: *First,* check the capacity of the *quart* container, using water, the funnel, and the standard measuring cup. Empty the quart container. Fill the liter container *to the brim.* Next, pouring carefully, and using the funnel, pour the water from the liter container into the quart container.

1. How do the capacities of the litre and the quart compare?
2. What are the dimensions of the litre container?
3. What else could the litre container be called to describe its volume?

E. MASS (I)

Equipment: One-kilogram mass (two identical litre containers; double-pan balance; small scraps of styrofoam (more than enough to fill a litre); sand (more than enough to fill a litre); cold water (as nearly 4° C as is practical); funnel, tray, towel, several sheets of newspaper; small amount of clay.

Procedure:
1. Place the empty litre containers on the double-pan balance, one on each pan. If the scales do not appear to be balanced, put a small bit of extra clay on the beam near the light end until they do balance.
2. Place the kilogram mass in one container.
3. Fill the other container with styrofoam.
4. How does the mass of the litre of styrofoam compare with the mass of the kilogram? (i.e., much ligher, much heavier, about the same, etc.)
5. Empty the styrofoam and *refill* the litre container with sand. (Work over a paper and use a funnel.)
6. How does the mass of the litre of sand compare with the mass of the kilogram, approximately? (See step 4.)
7. Empty the sand. Fill the litre container with cold water.
8. How does the mass of the litre of cold water compare with the mass of one kilogram? (See step 4.) What important relationship do you observe?

F: MASS (II)

Equipment: Two-pan balance; one-kilogram mass; three pounds of oleomargarine.

Procedure:
1. Balance sufficient oleomargarine against the kilogram mass.
2. One kilogram is equivalent to how many pounds?

G: MASS (III)

Equipment: Simple two-pan balance (homemade from cardboard if desired; pattern furnished. See Appendix II following this chapter.)

"Weights" one to five grams in mass (may be commercial and/or premeasured rectangular regions of file cards, bits of clay, short strips of lead, etc.). Suitable small objects for which to find mass (thumb tacks, piece of chalk, short pencil, paper clips, dried beans, sugar cubes, cuisenaire rods of various lengths, coins).

Procedure: Find the masses of three or more of the small objects.

H: EFFECT OF VOLUME CHANGE ON MASS

Equipment: Two-pan balance (a single beam balance could be used, but we used double-pan balances for all lab work); two identical containers (cottage cheese cartons used here); extra container for wet garbage; eggs; balloon; "weights" (or beans to use as counter balance, since the actual mass is not important, while the *conservation* of mass is); egg beater.

Procedure:

1. Separate an egg, carefully putting the white into one of the identical containers, and putting the yolk into the water-proof garbage container. Balance the container of egg whites against the other container and suitable "weights."
2. Remove the container of egg white from the balance; beat the egg white until it is light and fluffy.
3. Replace the beaten egg white on the balance.
4. Has the volume of the egg white changed? Has its density changed? Has its mass changed?
5. Balance the deflated balloon.
6. Blow up the balloon and replace on the balance.
7. Has the volume of the balloon changed? Has the density of the rubber changed? Has the mass of the balloon changed? Does the density of a particular fixed substance seem to affect its mass?

I: TEMPERATURE

Equipment: Containers (beakers, tin cans or small pitchers) of ice water and of boiling water; an induction heater or hot plate to boil the water (CAUTION: Do this in the presence of the teacher, and be very careful.); Celsius and Fahrenheit thermometers.

Procedure:

1. Read and record the temperature of both thermometers on a two-column chart labeled C° and F°
   a. in a sunny part of the room;
   b. in a shady part of the room;
   c. near the heater in the room;

d. far from the heater in the room;
e. near the floor in any part of the room, not near a heater;
f. as high in the air as you can reach, above the point you used in *e;*
g. in various locations outdoors if possible.

2. Immerse both thermometers in the ice water; allow to stand for about three minutes; read and record both.
3. Lay the two wet thermometers on a paper towel, wipe them dry, and allow them to rest for about three minutes.
4. Immerse both thermometers in the boiling water. Be careful of the steam, but hold both thermometers so the bulbs are well under the water but NOT touching the container or the heater. Read and record both.
5. Examine your charts to see how the Celsius and Fahrenheit temperatures generally compare.
   a. What is the temperature of freezing (ice) water on both?
   b. What is the temperature of boiling water on both?
   c. What interesting fact do you notice about the freezing and boiling points of water on the Celsius thermometer? Can you account for the fact that this thermometer has often been called a "Centigrade" thermometer?
   d. (Optional) Make a double "line" graph, using one color to show Fahrenheit temperatures and the other to show Celsius temperatures. What do you notice about the graphs in relation to each other?

J: GAMES

Equipment: Commercial and teacher-made games stressing prefixes, equivalent measures; estimating length. (See Appendix III following this chapter.)

Procedure: Find one or two or three opponents, and play one of the games. Evaluate its use for your classroom.

Evaluation: 1. Write a brief paragraph about the metric system, name and classify the basic units, and discuss the relationships among various units, the meaning of prefixes, and advantages of the use of the system.

2. What did you like about the laboratory experience?
3. What changes would you make in the metric workshop if you could?

CONCLUSION

In-service workshops for teachers have been held for a one-week period for college credit, at faculty meetings, at Saturday morning meetings, and at evening workshops for teachers in the area. Two stations were an outgrowth of the evaluation of the first workshop when it became evident that the important concept of a litre of *water* had not registered with many participants, even after lab experiences, filmstrips, films, and discussion. The meaningless phrase "a litre has a mass of one kilogram" appeared on too many evaluation sheets for comfort. The workshop was followed by a discussion sheet which attempted to clear up this point; it was decided to include another experiment in later workshops to stress the point. This was Station E in which litres of sand, water, and styrofoam bits were compared with a one-kilogram mass on a two-pan balance. Another experiment included in the GLCTM Workshop which was new, was Station H, in which the discovery was made that mass of a given portion of matter does not vary because of a change in density. This station will probably be kept in most of our future workshops, as the results were convincing, although the need for it was not as urgent as was the need for Station E.

A "Look and Listen" station for filmstrips and cassettes is desirable, and is used in the longer workshops when time allows.

Station C was suggested by an activity from *The Metric System,* (Addison Wesley, Menlo Park, California).

Evaluation remarks made by participants have been more than 95% enthusiastic. The deprecatory comments were mainly based on the brief period of time allotted in faculty meetings and 3-hour workshops, which was deplored but understood.

It was made clear in the introduction that there was time only to perform a few activities, to scan the others, and to share experiences in the discussion period that followed the laboratory experiences.

These brief experiences have been valuable to teachers in developing their own confidence and in promoting active involvement of their classes in becoming familiar with the elementary aspects of the metric system.

APPENDIX I

MODEL OF ONE CUBIC DECIMETRE (ONE LITRE)

Directions:

1. Cut a half-gallon milk carton 10 cm from the bottom.
2. Cover the sides of the carton with cm squared paper.
3. Cover the paper with clear Contact paper to protect your box.

This is a slightly inaccurate "litre," but it is a close enough approximation to convince pupils that:

(a) a litre holds slightly more than a quart;

(b) a litre of cold water (with mass of 1 kilogram) is about as heavy as 2.2 pounds;

(c) most important, that the litre, which is the metric unit of capacity is evolved sensibly from the metre, the metric unit of length, by being defined as 1 cubic decimetre (or 1000 cm^3).

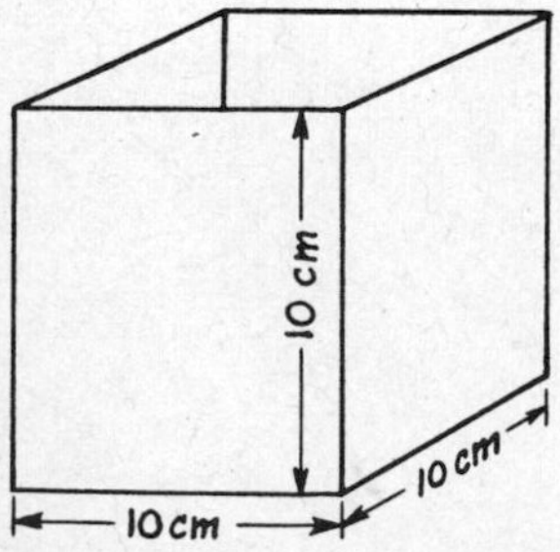

APPENDIX II

TAGBOARD BALANCE FOR FINDING SMALL MASSES

Directions:

1. Cut one copy of pattern A and two copies of pattern B from tagboard.
2. Punch holes at the indicated places at both ends of B and at the three corners of A.
3. Fold each B on line segments $\overline{MX}$, $\overline{NY}$, $\overline{MN}$, and $\overline{XY}$.
4. Using two metal paper fasteners, attach each pan B through holes a and b at the bottom of A.
5. Securely tape a pencil with a good eraser to a table top, extending about 6 cm from the table edge.
6. Insert a sturdy pin or long thumbtack through hole c in part A.
7. Using suitable "counterweights," the masses of such items as pencils, chalk, paper clips, sugar cubes, cuisenaire rods, etc., can be found to the nearest gram.

Examples:

Approximate masses: a nickel--five g; a paper clip--one g; a sugar cube--two g; a penny--three g; cuisenaire rods--one g per unit length.

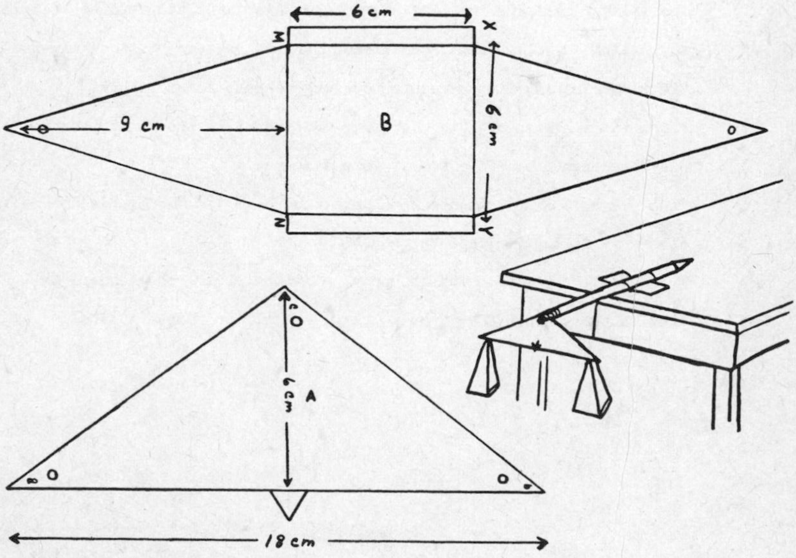

APPENDIX III

DOUBLE DECIMETRE (for 2 to 4 players)

Equipment: A set of forty cards, each printed with a line segment of length 1 cm, 2 cm, 3 cm . . . 10 cm. There are four cards for each of the 10 lengths.

Procedure: The leader places one card face down in front of each player including himself. He then places a second card face up in front of each player, including himself.

Each player looks at the line segments on both of his own cards, keeping the first card hidden from his opponents. Each person estimates (guesses) the combined lengths of the line segments on his two cards. If the lengths appear to total less than 20 centimetres (2 decimetres), he asks for more cards, face up, so that all can see them. When each person feels that he has nearly 20 cm total length, no more cards are given out.

Each player measures the line segments on his cards and reports the total length. Players may check each other's measures if they wish. The player having line segments totaling nearest 20 cm without being longer is the winner. His score for that round is the total length of his line segments. No one else gets a score.

If a player's line segments total exactly 20 cm (a "double decimetre") his score is doubled, and he gets 40 points.

The game is repeated, with the previous winner being the leader, as long as the players have time to play, or until a player's final score reaches 100 points.

SI BY SELF-INSTRUCTION AT A COMMUNITY COLLEGE

Robert L. Taylor

The North Campus of the Community College of Denver has a one hour course titled "The Metric System." This course is designed to give students in industrial occupations, health occupations, and general studies a basic understanding of the metric system as it is in use today.

The original course consisted of 10 one-hour class sessions in which the prefixes, symbols, and conversions within the two existing systems of measurement were taught. Many students were unable to register for this course or meet the class each week because of conflicting course or work schedules. To alleviate this situation a self-instructional package was developed and published by Multi-Media Publishing Incorporated, Denver, Colorado. The entire self-instructional program consists of the following lessons:

Lesson	Title	Content
A.	Basic Metric	Units, symbols and prefixes
B.	Basic Metric	Conversion within the metric system
1.	Length & Weight	A more advanced approach than Lessons A & B which covers units, symbols, prefixes, and inter-metric conversions
2.	Area, Cubic Volume & Capacity Volume	Teaches calculation of area and cubic volumes in metric units & the liter unit currently used for capacity volume

Robert L. Taylor, Ph. D. Instructor, Math and Science, Community College of Denver, North Campus.

3.	Length (English-Metric Conversions)	Teaches a basic method to make conversions from any length into any metric length desired
4.	Weight (English-Metric Conversions)	Teaches a basic method to make conversions from any weight into any metric weight desired
5.	Area (English-Metric Conversions)	Teaches a basic method to make conversions from any area unit into any metric area desired
6.	Cubic Volume (English-Metric Conversions)	Teaches a basic method to make conversions from any cubic volume unit into any metric cubic volume units desired
7.	Capacity Volume (English-Metric Conversions)	Teaches a basic method to make conversions from any capacity volume unit into any metric capacity volume units desired
8.	Temperature/ Calories/Density/ Specific Gravity	Celsius and Fahrenheit scales are compared and a method to convert is taught. The meaning and definitions of calories used as heat units and diet units are presented. Density and specific gravity are explained and a method to convert from one to the other is demonstrated.

Each of these lessons includes a workbook, which the student purchases from the bookstore, and the audio-visual presentation. With these three media the student "sees", "hears", and "responds" during the lesson. This *tri-media* program was planned for adult learning, yet uses a *skill* approach to minimize math operations. One time through for each lesson will average about 50 minutes. The audio-visual portion of each lesson "teaches" for about 10 to 15 minutes, then the student is directed to the workbook to practice and firmly grasp the subject matter presented. This involves the student in the learning process in a manner similar to classroom participation. When the student completes each lesson he takes a "Self-Test" and grades it with the answer sheets provided in the back of the workbook. He can then determine to his satisfaction that he has mastered the entire lesson. This also prepares him for the lesson evaluation given by the instructor. It is this last unit evaluation which is used to determine a grade for the student.

Lessons A and B are used as non-credit preparation for Lessons 1 through 8 or for students who need only the bare fundamentals of the metric system to apply to the science courses in which they are enrolled. Lessons 1 through 4 are currently in use as both a substitute and an aid for the classroom presentations at the Community College of Denver. Students who cannot meet the scheduled classes or do not feel they learn sufficiently from the class can and do use the tri-media material. Lessons 5 through 8 will soon become a part of a new one hour course for self-instruction at this college.

Students enroll for the self-instructional metric course in the same manner as for the regular course. They meet once with their instructor for instructions and are advised that they have many options open to them to complete the course. Both classroom and the tri-media learning methods may be used together in which instance their highest grade from either method is used for an evaluation and a grade in the course. Students with learning difficulties are often encouraged to use both methods. Students who have missed several sessions of the classroom presentations are encouraged to use the tri-media lessons to "catch up" to the other students. Students who are taking the course only by tri-media purchase lesson books and independently study their lesson at the Learning Material Center.

As of February 1975 only a small portion of the students enrolled for the metric course had taken it by self instruction. The self-instructional program was put into effect in the fall quarter of 1974 and since it was a new approach, student skepticism resulted in low participation. This has been corrected somewhat through better advertising and "word of mouth" recommendation on the part of former students.

A survey was conducted to compare the classroom and self-instruction methods and to sample student preferences. The data presented are somewhat limited since they are only for the fall quarter of 1974.

	Classroom Instruction	*Self-instruction*
Started course	40	12
Completed course	32 (80%)	11 (92%)
Grades	9 A's (28%)	10 A's (91%)
	12 B's (37.5%)	1 B (9%)
	11 C's or below (34.5%)	(No grades below B)

A "preference poll" was taken of those students who completed the self-instruction program and these results are listed below.

QUESTIONS	*RESPONSE*		
	No	*Yes*	*No answer*
Had metric before?	8	1	2
Prefer self-instruction to classroom?	0	9	2

Admittedly, the samples are small and no pretense is made that they prove the self-instructional method of teaching the metric system is better or more successful than the classroom. However, the combination of the two teaching methods offers a greater opportunity for the student to learn, and for the instructor to teach than either method would alone. The metric system seems well suited for the self-instructional mode of presentation since it consists of a listing of facts which are standard.

The use of a tri-media program for self instruction in the metric system at a community college is feasible. It adds another excellent method for a student to learn and master the subject. It also enables the instructor to direct the student to different methods of learning and increases the number of students who fulfill the course requirements in one quarter.

A FUNCTIONAL COURSE IN METRIC EDUCATION: A COMPETENCY-BASED APPROACH

Earl C. Leggette

INTRODUCTION

A need for a course in metrics grew out of ELED 564: *Current Trends in Teaching Mathematics in The Elementary School* taught by the author for Teacher Corps Interns in the fall quarter of the 1973–74 school year. This course (ELED 564) was competency-based and field-centered with a modular delivery system.

The module format adopted by the Teacher Corps Staff included Module Title, Rationale for Module, Behavioral Objectives, Pre-assessment, Learning Alternatives and Post-assessment.

One of the eight modules developed for studying current trends in teaching mathematics focused on "Items To Be Taught." This master module carried three submodules, one of which gave a general overview of metrics. At the conclusion of the course (in the student evaluation) came the recommendation that "with the current interest and growing availability of commercial material within the schools, metrics should be treated with a master module instead of a submodule under "Items To Be Taught."

Inservice teachers constituted the class for ELED 564 in the winter quarter, 1973–74. The revision of the course included a master module with two subs:

ELED 564 - 003 - L Modernized Metric System

ELED 564 - 003A - L Understanding The Metric System

ELED 564 - 003B - L Teaching The Metric System.

Student evaluation of this course suggested the possibility of a special workshop or even a course in metrics for teachers.

Earl C. Leggette, Ed. D. Associate Director of Teacher Corps and Associate Professor of Secondary Education (Mathematics) Jackson State University, Jackson, Mississippi.

As this evaluation was being administratively reviewed, a number of requests came from other teachers of the community for special training in metrics.

Early in the 1974 summer session, a decision was made to offer a fall mini-course in metrication for public school teachers. The course was to be a workshop in nature and offered through the Department of Continuing Education and the author was asked to teach it.

After several consultation sessions, the content of the course was delineated. The author then sequenced the material and decided on competency-based instruction as the functional instructional strategy.

In the fall quarter, 1974–75, from September 25 to October 9, Math 529: *Metrication For Public School Teachers* was offered.

The following is the basic information developed for the course:

-- for public information-- News Release
-- for student information-- course description and modules
-- for course data-- forms and summaries

COURSE DESCRIPTION

MATH 529 *Metrication For Public School Teachers* is designed to teach public school teachers the basic units of the national metric system as well as provide functional pedagogical techniques for these teachers to teach metric ideas to their students. The structure for the course will employ a competency-based module delivery system. The course will be a workshop in nature with a number of consultants. The primary focus will be on metric measurements for the following: length, area, mass (weight), volume, temperature. However, the complete SI (Le Système International d'Unités) system of measurement will be taught as an added feature depending upon individual demands and needs of participating students. Four graduate hours in mathematics will be given upon completion of the course.

There will be two modules:

1. Understanding the Metric System
2. Teaching the Metric System

Completion of both modules with at least 90% accuracy merits a grade of "A". Completion of both modules with at least 80% accuracy merits a grade of "B". Less than 80% accuracy on both modules merits a grade of "I".

Number: 529-001-L
Context: Current Trends In Instruction
Area: Measurement
Title: Understanding The Metric System
Target Population: Public School Teachers
Rationale: The first goal of the educational system is to bring virtually every individual to a level of literacy and competence which will enable him to become a productive member of his society. Measurement is one of the major strands in the educational system. The task of measuring and teaching measurement using common notions can be very complicated. The Metric System incorporates a less complicated venture. Therefore, metric ideas should be learned by teachers. The teachers will then be able to help their students explore metrics.
Objectives:

1. After the completion of this module, the participant will be able to select an appropriate measuring device and measure length, area, mass, liquid volume, and temperature in metric units with ninety percent accuracy. Acceptable performance will be ninety percent correct.
2. After the completion of this module, the participant will be able to convert from one system to the other: English to Metric, Metric to Metric. The units will be length, mass, volume, and temperature. Acceptable performance will be ninety percent correct.

Prerequisites: None
Pre-Assessment: Informal survey in an informal setting- "What do you know about metrics?"
Learning Alternatives:

1. Attend seminars by various consultants.
2. See a continuous nightly showing of films on metric in Room 318, SEB, from 6:00–7:40 p.m.
3. Attend a continuous nightly presentation of sound filmstrips on metric in Room 318, SEB, from 7:55–10:00 p.m.
4. Consult reserve book list in Jackson State University Library.

5. Attend lecture demonstrations by various consultants.
6. Schedule individual conference with professor.
7. Schedule group conference with professor.
8. Consult listings from Metric Reference List.
9. Check out from professor and review different metric games.
10. Check out from professor and review different activity kits and packages.
11. Free choice.

Post-Assessment: After completing the necessary learning alternatives for this module, complete the paper and pencil tests to be obtained from the instructor.

Number: 529-002-L
Content: Current Trends In Instruction
Area: Measurement
Title: Teaching The Metric System
Target Population: Public School Teachers

Rationale: Since today's teachers are responsible for preparing today's students for tomorrow's world, and tomorrow's world will also concern itself with quantity, the measurement concept should be taught in today's schools to today's students, even at the elementary level. Of the world population, a reported ninety-five percent are concerned with only four kinds of measurements in their daily lives-- length, volume, area and weight. Since tomorrow's world will be a Metric World, the idea of metric units should be taught for at least three of these four kinds of measurement-- metre for length, litre for volume and gram for weight.

Objectives: The participant will select a grade level and construct a module designed to teach the concepts-- metre, litre, and gram-- to an average student. The participant will submit to the professor a typed copy of the module developed plus all support material. The participant's module will be demonstrated by him in class in five minutes or less. The functionality of the module will be judged by members of the class and the professor using the attached critique sheet.

Prerequisites: Successful completion of module number: 529-001-L.

Pre-Assessment: The participant may opt out of this module by (1) or (2) below:

(1) if he can show evidence of having conducted at least one two-hour metric workshop for teachers using a competency-based approach to instruction with a module delivery system.

(2) if he submits to the professor at least ten (10) lesson plans with all support materials (posters, transparencies, etc.) that have been developed by him and used to teach metrics.

Learning Alternatives:

1. Hear lectures by various consultants and the professor.
2. Study number 529-002-Info.
3. Attend seminars by various consultants.
4. Attend lecture demonstrations by various consultants.
5. Schedule individual help conference with the professor.
6. Schedule group help conference with the professor.
7. Consult listings found in Metric Reference List.
8. Check out from professor and different metric games to review.
9. Check out from professor and various activity kits and packages to review.
10. Consult films list for time and place of showing.
11. Consult sound films list for time and place of showing.
12. Study: Thomas Nagel. *Competency-Based Instruction: A Strategy to Eliminate Failure.* Charles E. Merrill Co. 1972.
13. Free Choice.
14. Consult reserve book list in Jackson State University Library.

Post-Assessment: Submit the material completed for the objective with all support material. In addition an oral report will be given to the class on the module design on Wednesday, October 9, 1974.

CLASS SCHEDULE

Wednesday, September 25, 1974

6:00 - 7:40 p.m.

Measurement: Retrospect and Prospect--Part I

7:55 - 10:00 p.m.

Measurement: Retrospect and Prospect--Part II

Thursday, September 26, 1974

6:00 - 7:40 p.m.

Visual Approach to Teaching: What Is Metric? Why Metric?

7:55 - 10:00 p.m.

Work Session: Problem-Sharing
--Metric Made Easy--Part I

Friday, September 27, 1974

6:00 - 7:40 p.m.

Teaching Metric Ideas in Mathematics in the Elementary Schools
--A Practical Approach

7:55 - 10:00 p.m.

Work Session: Problem-Sharing
--Metric Made Easy--Part II

Monday, September 30, 1974

6:00 - 7:40 p.m.

Dimension Analysis

7:55 - 10:00 p.m.

Work Session: Problem-Solving
--Competency-Based Instruction: A Strategy to Eliminate Failure

Tuesday, October 1, 1974

6:00 - 7:40 p.m.

Metric Education in Secondary Education for the Sciences
--A Process Approach

7:55 - 10:00 p.m.

Work Session: Problem-Solving
--Metric Makes It Easy--Part I
--Planning Lessons

Wednesday, October 2, 1974

6:00 - 7:40 p.m.

A Metric Mississippi
--A Decision Whose Time Has Come

7:55 -10:00 p.m.

Work Session: Problem-Solving
--Metric Makes It Easy--Part II
--Planning Lessons

Thursday, October 3, 1974

6:00 - 7:40 p.m.

Practical Approach to Teaching Metric Education Today
--Teachers' Aids in the Classroom

7:55 - 10:00 p.m.

Work Session: Problem-Solving
--Preparing the Physical Environment for Learning
--Metric Games
--Planning Lessons
--Developing Materials

Friday, October 4, 1974

6:00 - 7:40 p.m.

Metric Education in Industrial Education

7:55 - 10:00 p.m.

Work Session: Problem-Solving
--Metric Education in Secondary Education for Mathematics--A Problem-Solving Approach
--Planning Lessons
--Developing Materials

Monday, October 7, 1974

6:00 - 7:40 p.m.

Teacher Aids in the Classroom

7:55 - 10:00 p.m.

The State of Metric Thinking
--Planning Lessons
--Developing Materials

Tuesday, October 8, 1974

6:00 - 7:40 p.m.

Metrication in Other Countries

7:55 - 10:00 p.m.

Work Session: Problem-Solving
--Planning Lessons
--Developing Materials
--Displays

Wednesday, October 9, 1974

6:00 - 7:00 p.m.

Metrication in Sports

7:00 - 10:00 p.m.

Displays and Demonstration of Material by Students

Number: 529-002-L Info
Competency-Based Instruction[1]
A Strategy To Eliminate Failure

Axiom I
In traditional programs time is held constant while achievement varies, while in a competency-based program achievement is held constant while time varies.

Axiom II
Traditional programs place greatest weight on entrance requirements, while competency-based programs place greatest stress on exit requirements.

Axiom III
If you want somebody to learn something, for heaven's sake tell him what it is!

Axiom IV
Competency-Based Instruction equals Criterion-Referenced Instruction plus Personalization of Instruction.

Competency-Based Teacher Education[2]
Progress, Problems, and Prospects

Most competency-based education programs employ a unit of learning called a *module*. The instructional module includes a set of activities intended to facilitate the learner's achievement of a specific objective or set of objectives.

Although the form or format of the module may vary with its setting and its objectives, most modules include five parts:

(1) Rationale: a clear statement explaining the importance and relevance of the objectives to be achieved.
(2) Objectives: stated in criterion referenced terms, specifying the considerations for successful completion.
(3) Pre-assessment: tests with the learner's competence in selected prerequisites and evaluates his present competence in meeting the objectives of the module. On the basis of this pre-assessment, the learner may opt out of the module, receive credit without further activities, or focus his efforts on areas of greatest need.
(4) Enabling Activities (Learning Alternatives): specify several procedures for attaining the competence specified by the module objectives. Individualization is promoted.

(5) Post-assessment: like the pre-assessment, measures competency in meeting the module objectives. Unsuccessful performance usually leads to recycling through the optional activities (a closed loop). Modules also include feedback mechanisms by which students are kept informed of their performance and progress.

CHAPTER NOTES

1. Thomas Nagel and Paul Richman, *Competency-Based Instruction: A Strategy to Eliminate Failure* (Columbus, Ohio: Charles E. Merrill Publishers, 1972.
2. Robert W. Houston and Robert B. Howsam, *Competency-Based Teacher Education: Progress, Problems, and Prospects* (Chicago: Science Research Associates, 1972.)

METRIC REFERENCE LIST

Ambler, E. "Measurement Standards, Physical Constants, and Science Teaching," *Science Teacher* (November 1971): 63–71.

Barbrow, L. E. "Metrication in Business and Industry," *Business Education Forum* (December 1973): 8–9.

Batcher, Olive M., and Young, Louise A. "Metrication and the Home Economist," *Journal of Home Economics* (February 1974): 28-31.

Branscomb, Lewis M. "The U.S. Metric Study," *Science Teacher* (November 1971): 58–62.

Bright, George W. "Metric, Students, and You!" *Instructor* (October 1973): 59–66.

Bright, George W., and Jones, C. "Teaching Children to Think Metric," *Today's Education* (April 1973): 16–19.

Caravella, Joseph R. "Metrication Acitivities in Education," *Business Education Forum* (December 1973): 14–16.

Cortese, Carole E. *Metric Measurement.* (New York: American Book Co., 1970.

Cortright, R. W. "Adult Education and the Metric System," *Adult Leadership* (November 1971): 190.

DeSimone, Daniel V. *A Metric America: A Decision Whose Time Has Come.* Bulletin C13.10:345. Washington, D.C.: Government Printing Office, 1971.

Donovan, Frank R. *Prepare Now for a Metric Future.* New York: Weybright and Talley, 1970.

Edson, Lee. "New Dimensions for Practically Everything: Metrication," *American Education* (April 1972): 10–14.

Gaucher, Claire, and Perry, Sophia. "What Impact Will Metrication Have on Home Economics?" *Forecast to Home Economics* (March 1974): 14–16, 72.

Hallerberg, Arthur E. "The Metric System: Past, Present,-- Future?" *Arithmetic Teacher* (April 1973): 247–55.

Helgren, Fred J. *Metric Supplement to Science and Mathematics.* Oak Lawn, Ill.: Ideal School Supply Company.

Hopkins, Robert A. *The International (SI) Metric System and How It Works.* Tarzana, Calif.: Polymetric Services, 1974.

Howell, Daisy. *Activities For Teaching Mathematics To Low Achievers.* Jackson: University Press of Mississippi, 1974.

Huffman, Harry. "How Business Teachers Should Plan for Metrication," *Business Education Forum* (December 1973): 17–19.

Johnson, J. T. "A Study of Weights and Measures: Another View," *Mathematics Teacher* (May 1944): 219–21.

Joly, R.W. "Changing to the Metric System," *NASSP Bulletin* (November 1972): 47–59.

Jones, Philip G. "Metrics: Your Schools Will Be Teaching It and You'll Be Living It -- Very, Very Soon," *American School Board Journal* (July 1973): 21–26.

Kempf, Albert F., and Richards, Thomas J. *Exploring The Metric System.* Atlanta: Laidlaw Brothers, 1973.

Kendig, Frank. "Coming of the Metric System," *Saturday Review* (November 25, 1972): 40–44.

King, Irv, and Whitman, Nancy. "Going Metric in Hawaii," *Arithmetic Teacher* (April 1973): 258–60.

Laycock, Mary, and Watson, Gene. *The Fabric of Mathematics.* Hayward, Calif: Activity Resources Company, 1971.

Leggette, Earl C. *The Effect of A Structured Problem-Solving Process on The Problem-Solving Ability of Capable But Poorly Prepared College Freshmen In Mathematics.* (Doctoral dissertation, Rutgers University) 1973. DAI 34.

Metric Association. Metric Supplement to Science and Mathematics. Boulder, Colo.: The Association, 1971.

Metric Implementation Committee of the NCTM. "Metric: Not If, But How," *Arithmetic Teacher* (May 1974): 366–69.

Murphy, M.O., and Pelzin, M. A. "Descriptive Analysis of the Teaching of the Metric System in the Secondary Schools," *Science Education* (February 1969): 89–91.

Nagel, Thomas S., and Richman, Paul. *Competency-Based Instruction: A Strategy to Eliminate Failure.* Columbus, Ohio: Charles E. Merrill Co., 1972.

National Bureau of Standards. "A Metric America" and "A Metric Conversion Case Study," *Technical News Bulletin* (September 1971) Washington, D. C.: Government Printing Office.

National Bureau of Standards. *Brief History of Measurement Systems* (chart of modernized metric system). Washington, D.C.: Government Printing Office.

National Council of Teachers of Mathematics. *Arithmetic Teacher* (April 1973) issue devoted to metric; some metric articles in May 1973.

National Council of Teachers of Mathematics. *A Metric Handbook for Teachers.* Jon L. Higgins, ed. Reston: The Council, 1974.

National Council of Teachers of Mathematics. Measurement, *Topics In Mathematics for Elementary School Teachers* #15, Washington, D.C., 1968.

National Council of Teachers of Mathematics. Instructional Aids In Mathematics. *Thirty-fourth Yearbook.* Washington, D.C.: NCTM, 1973.

National Education Association. "Metric System Ahead," *Research Bulletin* (December 1971): 100–12.

National Education Association. "Think Metric," *Research Bulletin* (May 1971): 39–42.

Odom, Jeffrey C. "The Current Status of Metric Conversion," *Business Education Forum* (December 1973): 5–7.

Odom, Jeffrey V. "The Effect of Metrication on the Consumer," *Business Education Forum* (December 1973): 10–13.

Page, Chester H., and Vigoureux, Paul. *The International System of Units (SI)."* National Bureau of Standards publication no. SP-330. Washington, D.C.: Government Printing Office, April 1972.

Ploutz, Paul F. *The Metric System, A Programmed Approach.* Columbus, Ohio: Charles E. Merrill, 1972.

Prakken Publications. *School Shop* (April 1974) issue devoted to metrics.

Ritchie-Calder, Lord. "Conversion to the Metric System," *Scientific American* (July 1970): 17–25.

Rucker, Isabelle P. "The Metric System in Junior High School," *Mathematics Teacher* (December 1958): 621–23.

Schools Council. *"Metres, Litres and Grams."* London: Schools Council Publications; New York: Citation Press, 1971.

Schreiber, Edwin W. "Significant Facts in the History of the Metric System for Teachers of Junior and Senior High School Mathematics," *Mathematics Teacher* (November 1929): 373–81.

Smith, Eugene P. "The Metric System: Effects on Teaching Mathmatics," *NCTM Newsletter* (March 1974): 1.

Souder, Wilmer. "The Metric System: Its Relation to Mathematics and Industry," *Mathematics Teacher* (September 1920): 25–35.

Strong, E. "Meaningful Metric," *School Science and Mathematics* (May 1964): 421–22.

Vervoort, Gerardus. "Inching Our Way toward the Metric System," *Mathematics Teacher* (April 1973): 297–302.

Williams, Elizabeth. "Metrication in Britain," *Arithmetic Teacher* (April 1973): 261–64.

Yorke, Gertrude Cushing. "A Study of Weights and Measures," *Mathematics Teacher* (March 1944): 125–28.

ADDITIONAL SOURCES OF METRIC INFORMATION IN THE UNITED STATES

American National Metric Council, 1625 Massachusetts Avenue, NW, Washington, D.C. 20036.

American National Standards Institute, 1430 Broadway, New York, New York 10018.

Metric Association, Sugarloaf Star Route, Boulder, Colorado 80302.

Metrication Task Group, George C. Marshall Space Flight Center, Huntsville, Alabama 35812.

Metric Information Office, National Bureau of Standards, Washington, D.C. 20234.

National Council of Teachers of Mathematics, 1906 Association Drive, Reston, Virginia 22091.

National Science Teachers Association, 1742 Connecticut Avenue, NW, Washington, D.C. 20009.

IN OTHER COUNTRIES

Metric Advisory Board, P.O. Box 10243, The Terrace, Wellington, New Zealand.

Metrication Board, 22 Kingsway, London WC2B 6LE, England.

Metric Commission, 320 Queen Street, Ottawa KIA OH5 Canada.

Metric Conversion Board, 18–24 Chandos Street, St. Leonards, N.S.W. 2065 Australia.

South African Bureau of Standards, Private Bag X191, Pretoria, South Africa.

STUDENT EVALUATION FOR MATH 529: METRICATION

FOR PUBLIC SCHOOL TEACHERS

Answer all questions to the best of your ability. Do not sign form.

1. Please rate the overall quality of instruction in this course:

   _22_ Excellent _18_ Good ___ Average ___ Fair

2. For you, the material was presented at a pace which was:

   _2_ Much too fast _14_ Somewhat fast _22_ About right

   _2_ Somewhat slow ___ Much too slow

3. Rate the content of the course with regard to your present and future needs:

   _16_ Excellent _22_ Good _2_ Average ___ Fair ___ Poor

4. How has this course affected your interest in the subject area?

   _34_ Greatly increased _6_ Somewhat increased ___ Unaffected

   ___ Somewhat decreased ___ Greatly decreased

5. Considering the background you hope to attain in this subject area, the focus of this course is:

   ___ Much too narrow _4_ Somewhat narrow _26_ About right

   _8_ Somewhat broad _2_ Much too broad ___ Irrelevant

6. I made use of conference hours _26_ Yes _12_ No

7. I made use of library _24_ Yes _16_ No

8. I made use of Curriculum Lab _19_ Yes _21_ No

9. What did you enjoy most about this course?

   * (see attachment for different responses)

10. What is your major complaint with regard to this course?

    ** (see attachment for different responses)

*9 Coded Response to item 9

Variety of consultants for different areas	13
Competency-based instruction with modules delivery system	7
Learning activities	5
Preparing instructional materials	5
Problem solving	5
Informal instruction	3

**10 Coded Response to item 10

None. I got my money's worth!	28
Not enough time	6
Test was not necessary	2
Did not have representatives for book and school supply companies	1
Hours too long	3

TEACHER TRAINING IN CONSUMER METRICS

Clinita A. Ford

A teacher training workshop was conducted in 1974 for home economics teachers in Florida to gain experiences in the fundamentals of metrics. Primarily the workshop was designed to benefit secondary (pre-service and in-service) home economics teachers.

The workshop was funded by the Home Economics Section of the Department of Education, State of Florida.

The objectives of the workshop were to:

1. Acquire a general knowledge of basic metrics.
2. Gain fundamental skills in the use of metrics.
3. Relate basic metrics to the teaching of home economics.
4. Acquire familiarity with authentic resources in metrics.
5. Recognize multi-approaches to metrics through utilization of interdisciplinary/interdepartmental personnel.
6. Experience an involvement in translating metric fundamentals to the development of instructional materials for the teaching of home economics.

Instruction was handled by an interdisciplinary team of university faculty from areas of home economics and from mathematics. Home economics faculty involved with the workshop represented the areas of food and nutrition, home and family, furnishings and equipment and home economics education.

The workshop experiences were developed to be relevant to all aspects of home economics . Emphasis was on those measurements basic to the routine daily activities of the consumer.

Clinita Arnsby Ford, Ph.D., R.D. Professor of Home Economics, Director of Consumer Metrics, Florida A & M University, Tallahassee, Florida.

The workshop was conducted in a manner to encourage the participants to "think metrics" not "convert to metrics." The conversion crutches were not permissable in the workshop.

The participants explored approaches to the incorporation of metrics into high school class activities. Limited time was spent with this because a 1975 sequential workshop is planned for in-depth study of methodology, program design and curriculum development with the incorporation of metrics.

A unique feature of the workshop was the inclusion of senior high school students as active participants in the workshop. They served as youth monitors, the special sensors of the workshop. Since the overall purpose of the workshop was teacher training, it was considered appropriate to have secondary students involved in all activities. The youth monitors were full time participants in all workshop activities. They could react to our goals, directions, and activities as to potential impact on the high school student.

All participants had opportunities to engage in developmental activities. These stimulated creativity in the expression of concepts of metrics in home economics. The participants wrote radio and television scripts, did role-playing, developed a variety of instructional aids, designed puzzles and games and created an array of other metric related materials.

A cadre of select consultants enriched the total workshop experience. Those invited to serve as consultants each had a special expertise needed in the workshop and each had established national recognition in their respective fields.

The basic operational structure of the workshop scheduled consultants as the first activity of the morning followed by audiovisual reviews. In the afternoons basic instruction in metrics and group activities were scheduled.

The consultants handled the following topics:

"Visual approaches to metric education."
"Humanizing metrics: The psychology of change."
"An educational challenge: Metrics in home economics."
"Metrics for the home economist."
"Status and outlook of metric education with implications for home economics."
"Consumer metrics and its application in home economics programs."
"Metrics and life styles."

"Legislation and implementation of metrics: Impact on the consumer."
"Metrics in the marketplace."
"A perspective of metric concepts, standards and practices."
"Metrics: New meanings in clothing and textiles."

A variety of metric materials and instruments were available for participant use. These included:

Slides, Films and Filmstrips
Transparencies
Cassette Programs
Programmed Materials
Books, Bulletins, Pamphlets
Charts and Posters
Instructional Kits
Games
Household Metric Instruments

A "resource review", during a half-day in the second week of the workshop, permitted the participants the opportunity to handle, examine and use numerous items on display.

Instruction in basic metrics included seven (7) measurements:

Linear	Force
Volume	Square
Mass	Cubic
Temperature	

Participants prepared brief lists of suggested metric activities based on the seven (7) basic metric measurements for given subject matter in home economics. The purpose was for the participants to get a feel for adapting metric principles for their professional work.

Group activities were scheduled during the afternoon. These were structured for participants to experience several techniques and approaches to interpreting metrics in high school home economics classes. The participants designed games and puzzles, prepared scripts for skits and television shows and wrote short stories, poetry and advertisements in metrics.

A treasure hunt was a class activity conducted in a home. Participants were given challenging activity questions which could be answered only through active involvement. Various objects in the house were described metrically and were to be found from said description(s). Other questions were based on recognizing and locating in the home an unknown from metric data given.

Each participant prepared an instructional aid for the culminating project. The instructional aids were designed and created by the participants for use in classes they teach. These projects included charts, posters, transparencies, songs, games, mini-bulletin boards, mock-ups learning activity packages.

The instrument for workshop evaluation was distributed the first day of workshop. In this way the evaluative process was daily, eliminating the influence of recency of activities or memory. The evaluative items were:

Consultant presentations
Instructional activities
Instructional content
Support services (adequacy)
Strengths and weaknesses (identification)

Participants were asked, too, for an evaluation of the value of the workshop to them. These criterion factors were:

Contribution to professional growth
Expansion of professional knowledge
Value to improving effectiveness
Acquisition of new/useful ideas
Interest appeal

The participants were evaluated according to:

1. Quality of participation
2. Level and quality of productivity
3. Degree of growth and level of performance as measured by pretesting, daily testing and post testing.

The Consumer Metrics workshop has been a seed project, germinating motivation, interest, and a spiral of related educational activities. The Division of Consumer Science and Home Economics at Florida A & M University expects to further explore consumer metric education and continue to serve professional home economists as a source of training for keeping up-to-date.

SURVEY SHOWS NEED FOR METRIC WORKSHOPS FOR ELEMENTARY TEACHERS

John H. Trent

In order to determine whether or not there was a need for metric education workshops for elementary teachers in Nevada, questionnaires were sent to a random sample of this population in January, 1975. Of the sample, 71% returned the questionnaire.

The questions asked regarding the need for metric education and the responses received are shown in Table I below:

Table I

Need for Metric Workshop Questionnaire

1. Have you had a college course in which the metric system was taught or used? Yes 18% No 82%
2. Do you feel qualified to teach arithmetic (or science) courses in which the metric system is taught or used? Yes 20.4% No 79.6%
3. Did you know that in 1974 Congress passed a law stating that "education systems should be encouraged to provide metric education for students?" Yes 85% No 15%
4. Did you know that the Nevada State Textbook Commission has recommended that all textbooks adopted after January 1, 1976, have the metric system as the primary system of measurement? Yes 62.7% No 37.3%
5. How adequately prepared in the metric system are students when they commence the school year in your class?

0.0%	5.5%	24.2%	70.3%
Very well prepared	Fairly well prepared	Inadequately prepared	No preparation

John H. Trent Professor of Mathematics and Science Education, University of Nevada, Reno, Nevada.

6. How much are you now teaching the metric system to your students?
   3.1% ____ 56.6% ____ 40.3% ____
   A lot A little None at all
7. If a federally funded inservice course in metric education were offered by the University of Nevada, Reno, would you attend it?
   a. if it were offered in your county
      Yes 91.1% No 8.9%
   b. if it were offered on the University of Nevada, Reno campus
      Yes 53.3% No 46.9%
8. How great is the need for placing more emphasis on the metric system in elementary mathematics classes?
   62.4% ____ 36.0% ____ 1.6% ____
   Very great Needed somewhat No need
9. Do you feel that adequate guidelines, course outlines, and materials on the metric system are available to you for satisfactorily teaching the metric system to your students? Yes 24.6% No 75.4%

The data from Table I shows that a high percentage of Nevada elementary teachers:

(a) had not had a course in which the metric system was taught or used,
(b) do not feel qualified to teach arithmetic or science courses in which the metric system is used or taught,
(c) do not feel that their students are adequately prepared in the metric system, and
(d) do not feel that adequate guidelines, course outlines, and materials on the metric system are available to them for satisfactorily teaching the metric system to their students.

Even though these teachers are aware that both Congress and the Nevada Textbook Commission have recommended metric education for students and feel that the need for placing more emphasis on the metric system is great, over 95% of them are teaching little or none of the metric system to their students.

From Table I, the inference may be drawn that there is a great need for in-service workshops on the metric system for Nevada elementary teachers. The survey further indicates that high percentages of the teachers would attend these workshops, especially if they were held in their own counties.

In order to determine the present knowledge and ability of these teachers to use the metric system, another questionnaire was administered to the same sample at the same time. The questions asked and the responses received are shown in Table II below:

Table II

Knowledge of Metric System Questionnaire

1. The average lineman in the National Football League weighs:
   ____a. 15 kilograms
   ____b. 115 kilograms
   ____c. 225 kilograms
   ____d. 325 kilograms
   ____e. 425 kilograms

   (b)
   Correct Response 42.5%
   Incorrect Response 57.5%

2. The height of the average American male is:
   ____a. 1.85 centimeters
   ____b. .185 meters
   ____c. 1.85 meters
   ____d. 18.5 meters
   ____e. 18.5 centimeters

   (c)
   Correct Response 66.1%
   Incorrect Response 33.9%

3. The average American car gasoline tank holds:
   ____a. 80 liters
   ____b. 180 liters
   ____c. 8 liters
   ____d. .8 liters
   ____e. 1800 liters

   (a)
   Correct Response 55.2%
   Incorrect Response 44.8%

4. Match the numbers below to the letters:

____a. meter	1. .001 meter
____b. centimeter	2. .01 meter
____c. millimeter	3. 39.37 inches
____d. kilometer	4. .1 meter
____e. decimeter	5. 1000 meters

a. =_3_, b. =_2_, c. =_1_
d. =_5_, e. =_4_
Correct Response 59.4%
Incorrect Response 40.6%

5. The temperature on a hot day in central Nevada is about:
   ____a. 17° Celsius
   ____b. 212° Celsius
   ____c. 37° Celsius
   ____d. 100° Celsius
   ____e. 57° Celsius

(c)
Correct Response 18.1%
Incorrect Response 81.9%

6. What does MKS stand for?

(Meter, Kilogram, Second)
Correct Response 3.1%
Incorrect Response 96.9%

7. What does SI stand for?

(The International System of Units)
Correct Response 7.1%
Incorrect Response 92.9%

Between 42.5% and 66.1% of the teachers were able to correctly respond to questions relating to meters, kilograms, and liters. However only 18.1% of them correctly responded to questions related to Celsius temperature. Further, over 90% did not know the meaning of SI or MKS.

No doubt some teachers obtained some assistance from others and used conversion tables. Thus it is quite likely that a majority of the teachers are unable to "Think Metric."

In summary, it appears that the data from Table II further substantiates the need for metric education for Nevada elementary teachers.

The population of interest of this study was Nevada elementary teachers and hence the findings of this project cannot be generalized beyond this population. However, the results may cautiously be applied to groups of teachers similar to Nevada elementary teachers. With this caution, the overall conclusion of this study is that there is a great need for metric education for Nevada elementary teachers, and for other similar groups of teachers.

RESEARCH IN TEACHING THE METRIC SYSTEM

Robert M. Todd, Larry J. Weber, Thomas G. Teates and Maurice Esham

Prior to 1967 little research specifically related to the metric system had been published. The February, 1969, *Journal of Educational Research* article by Polzin and Murphy[1] titled "A Review of Research Studies on the Teaching of the Metric System" listed only 13 references.

Of the 13 references, two were early studies (1929, 1930) on knowledge of units and metric usage in chemistry texts; four others were based on pre-World War II studies and continued the argument of whether we should change to metric; three were studies of ability to estimate (poor for 5th and 6th grade students and somewhat better for college juniors and in-service teachers—but still poor); one compared success on measurement problems by traditional vs School Mathematics Study Group students (SMSG won); Roth's thesis "The Desirability and Practicality of Adoption of the Metric System in the U.S." [1964 Congressional Record], a one page item in *Science News* (January 19, 1967) and Polzin's analysis of the teaching of metrics in Washoe Co., Nevada, completed the list.

While people were still arguing whether to teach metric, there was little effort to find ways to teach it better.

Robert M. Todd, Ed. D., Associate Professor, Mathematics and Education, Virginia Polytechnic Institute and State University, Blacksburg, Virginia.

Larry J. Weber, Ed. D., Associate Professor, Virginia Polytechnic Institute and State University, Blacksburg, Virginia.

Thomas G. Teates, Ph. D., Associate Professor, Science Education, Virginia Polytechnic Institute and State University, Blacksburg, Virginia.

Maurice Esham, M.S., Assistant Professor, Morehead State University, Morehead, Kentucky.

RECENT STUDIES

Scott[2] found children's problem solving was no worse on measurement (continuous variable) problems than on non-measurement (discrete variable) problems; one of his recommendations was for school programs to stop avoiding measurement.

McFee[3] developed and validated a test of metric competence and compared 7th grade science students' success on the test when taught metric measures without conversion to English or when taught metric measures with simultaneous conversion to English (no significant difference). A control group (n:50) showed no gain between pre- and post-tests; thus gains were not the result of practice.

Exum[4] compared the effectiveness of the Metric Assn. booklet *Metric Supplement to Science and Mathematics* with usual instruction for undergraduate non-science majors in a science course using McFee's test as a measure. The groups using the Metric Supplement gained more than the control group from pre- to post-test. (Control group gained 15.71.)

Dod,[5] in a similar study, compared the Metric Assn. booklet's effectiveness to that of the Earth Science Curriculum Project's for teaching metric units to 8th grade students. The booklet students scored better on the McFee test but were noncommittal on an attitude scale on metrics. Dod also used the McFee test with general science teachers (mean of 28.10 on 40 item test for booklet group, 30.60 for ESCP group).

Bargman[6] set up an instructional program on the metric system of units based on measurement activities and taught it to 201 children in grades 3 through 6. He concluded: third grade children can learn metric units (and approximate sizes) of length, liquid volume (graduated cylinders) and mass/weight, and how to measure and make simple conversions between units using whole numbers.

Fourth grade children can learn area and cubic volume in addition to the above.

Fifth graders can do the above using decimals.

Sixth graders can do all the above.

Bargman concluded, "Adoption of the Metric System as the primary system of measurement in this country could result in curricular changes in elementary school mathematics, such as earlier introduction of decimal fractions and corresponding instruction in place value. If this is true, certain phases of the metric system could possibly be taught earlier than was implied by the findings of this study."

Esham,[7] 1974, conducted a study to compare the effectiveness of the *Metric Supplement* to an activity-oriented *Metric Measurement Guide* which he had prepared.

The entire class load, four seventh-grade and two sixth-grade classes, of an elementary mathematics teacher, was used in the study. Two classes, one a high-ability group and the other a low-ability group, were split and the students were assigned randomly to one of the two treatments; the other classes were assigned to a treatment as intact units. The total number of students was 134; instructional time, excluding testing, was five periods of 45 minutes each.

Students were pre-tested and post-tested on a Criterion Referenced (CR) Test on the metric system and also post-tested on the McFee test. Mean scores on the post test and gain scores on the CR Test were higher for students using the activity-based guide than for students using the Metric Supplement.

Results

The results of the criterion referenced and McFee tests are shown in Table I.

Table I

Class	Treatment	Criterion referenced pre-test	Criterion referenced post-test	Gain	McFee Test Total
1	Metric Supp.	9.62	11.91	2.29	15.07
2	Metric Guide	9.30	13.70	4.40	15.13
3a	Metric Guide	8.50	10.60	2.10	14.22
3b	Metric Supp.	6.43	8.85	2.42	11.16
4a	Metric Guide	11.83	15.23	3.40	23.42
4b	Metric Supp.	11.14	14.71	3.57	21.00
5	Metric Guide	10.39	15.77	5.38	17.00
6	Metric Supp.	8.50	12.10	3.60	15.10

Table II shows the values obtained by using the t-test in comparing the gain of different treatment groups.

Table II

	t value	degrees of freedom
Between classes 1 & 2	1.51	41 *
Between classes 5 & 6	2.45	46 **
Between group 3a & 3b	.71	14
Between group 4a & 4b	.30	25

*—significant at .10 level (one tail test)
**—significant at .01 level

Esham (1974) used the *Metric Measurement Guide* with 93 sixth and 168 fifth grade students in three elementary schools in southwestern Virginia.

The first period was devoted to pretesting (30 minutes) and the last fifteen minutes of the period were devoted to an introduction to the metric system by the classroom teacher.

Following the introduction, each student received a copy of the *Metric Measurement Guide,* a two-meter tape measure and centimeter square graph paper. Cuisenaire rods, bathroom scales graduated in kilograms, balances with metric weights and graduated cylinders were available for the students' use at all times during the period of instruction. The period of instruction consisted of five days (one 45-minute period per day). The classroom teacher was present during this time to answer questions when asked or to give additional directions when needed.

A twenty-item criterion referenced test, which was the same test as the pretest, was administered following the fifth day of instruction.

Results of the pre-test and post-test are illustrated in Table III. Results of the pre-test—post-test analysis reveal a significant

Table III

	Pre-test		Post-test			
	X	SD	X	SD	Gain	N
Fifth Grade	6.79	2.31	8.86	2.89	2.07	168
Sixth Grade	7.01	2.69	10.49	3.12	3.48	93

gain for both grades, although the sixth grade exhibited a considerably higher gain when compared with the fifth.

Sweetser and Todd used Esham's revised *Metric Measurement Guide* in self-directed activities for pre-service elementary education majors at the end of the junior year. The students spent four hours taking the McFee test as pre-test, working through the Metric Measurement Guide activities as part of their assigned math-science methods laboratory, and taking the McFee test as a post-test. Two weeks later the McFee test was again administered (unannounced) as a retention test.

McFee Test Scores of 40 pre-service elementary teachers, are shown in in Table IV.

Table IV

	Pre		Post		Retention	
	Mean	SD	Mean	SD	Mean	SD
General Proficiency	8.80	3.452	12.38	4.933	15.50	4.238
Intuitive	8.24	3.413	12.79	3.391	12.95	2.96
Total	17.04		25.17		28.45	

Measurement activities between the post-test and retention test were two laboratory exercises:

1) a determination of "Pi" by constructing circles and measuring the circumference (10m radius and 10 ft radius).
2) a mapping exercise using hypsometer and stadia devices.

The results on the retention test were pleasantly surprising; the gains from pre- to post-test were not lost through forgetting; instead, the students continued to learn. The score of 28.45 is impressive when compared to the scores of Dod's science teachers (28.10 mean on the same McFee test for teachers using the booklet, 30.60 for ESCP teachers). Esham is currently studying how well pre- and in-service teachers learn the metric system and the extent of attitude changes that occur during the instructional period. He is using the McFee test, the Semantic Differential scale as reported in Dod's study and an attitude scale he developed for the study. Instructional time was less than four hours.

Analysis is incomplete but preliminary results show:

Inservice teachers gained from pre-test mean of 20.17 to post-test mean of 28.77 on the McFee test. n = 61

Pre-service teachers in elementary education gained from pre-test mean of 16.45 to a post-test mean 27.13 on McFee's test. n = 72

SUMMARY

Previous research at University of Southern Mississippi showed the *Metric Supplement to Science and Mathematics* was successful in increasing scores on the McFee test, that science teachers using ESCP had a mean score of 30.60 while those using the booklet had a mean score of 28.10, and that 8th grade students using ESCP materials or the Supplement had low scores on the McFee test.

Recent research at Virginia Tech shows that:

1) Sixth and seventh grade students scored greater gains and preferred Esham's materials (based partly on the Metric Supplement but involving more activity) to the Supplement.
2) Pre-service teachers make substantial gains in knowledge of the metric system with less than four hours instruction and that further measurement activity seems to result in further gains.
3) In-service teachers also show significant gains with less than four hours of measurement self instruction.

CHAPTER NOTES

1. Mary O. Murphy and Maxine A. Polzin, "A Review of Research Studies on the Teaching of the Metric System," *The Journal of Educational Research* 62 (1969): 267–270.

2. Lloyd Scott, "A Study of the Case for Measurement in Elementary School Mathematics," *School Science and Mathematics* 66: 714–722.

3. E.E. McFee, "The Relative Merits of Two Methodologies of Teaching the Metric System to Seventh Grade Science Students" (Doctoral diss., Indiana University, 1967).

4. Kenith Gene Exum, "Evaluation of a Metric Booklet as a Supplement to Teaching the Metric System to Undergraduate Non-Science Majors" (Doctoral diss., University of Southern Mississippi, 1972).

5. Bruce Dod, "A Comparison of Two Methods of Teaching the Metric System to Eighth Grade Science Students" (Doctoral diss., University of Southern Mississippi, 1973).

6. Theodore John Bargman, "An Investigation of Elementary School Grade Levels Appropriate for Teaching the Metric System" (diss., Northwestern University, 1972).

7. Maurice Esham, "A Pilot Study in Methods of Metric Education Conducted at Margaret Beeks Elementary School" (Unpublished paper, Morehead State University, 1974).

CRITERION REFERENCED TESTING IN METRIC EDUCATION

Larry J. Weber, Robert Todd, Thomas Teates

There is nothing new about educational innovation. In the history of American education there have been scores of approaches designed purporting to be the answer to the learning problems of our children. It would be folly to mention some of them. Others have made an impact. However, one fact remains clear. While the efforts of educational innovators have not been without merit, it could be said that none has measured up to the total expectations of its creator. Fads come and go; we take what we can from each for our benefit and for those we teach; we alter them, and create new fads. Such is the process in which we engage.

In recent years one of the more popular developments in education has been the criterion test movement. A criterion referenced test is defined as "one that is deliberately constructed so as to yield measurements that are directly interpretable in terms of specified performance standards."[1] Several factors gave rise to the criterion testing movement. Those that have been prominently mentioned are the emphasis in recent years on educational accountability and individualized instruction.

Traditionally, educational progress of students and school systems was measured by means of standardized achievement tests. Disenchantment with such tests because of the following factors has resulted in a decreased use of them in schools throughout the U.S. First, the fact that they were culturally biased was seen as a causative factor for poor performance on them by minority group members. Secondly, standardized tests

Larry J. Weber, Ed. D., Robert Todd, Ed. D., and Thomas Teates, Ph. D. Associate Professors of Education, Virginia Polytechnic Institute and State University, Blacksburg, Virginia.

did not adequately measure many of the objectives of individual school systems. Such tests were constructed for widespread use and, consequently, reflected the goals of instruction of given schools in a general sense. Finally standardized test results were of limited value for helping individual students with their problems.

The deficiencies of standardized tests for purposes of instruction coupled with the aforementioned stress on accountability gave rise to the necessity of developing other types of instruments for determining educational progress of students. Such tests needed to possess the characteristic of providing a standard to which the individual himself could be compared, a standard which was independent of the performance of other individuals. They also needed to reflect the goals of instruction of educational programs for individual students and individual school systems. Criterion referenced tests, as they came to be called were constructed to meet these criteria.

Gronlund[2] lists several principles of criterion referenced testing that are helpful in the construction of such instruments. Included in them are the need to define and delimit a domain of learning tasks, the requirement that objectives be defined in behavioral terms, the necessity of specifying performance standards, the need to sample adequately student performance and the need for agreement between test items and objectives. During the past years efforts to develop criterion test materials for facets of metric education have been conducted by the authors of this paper and others[3]. The content of this paper will focus on problems associated with construction of criterion referenced tests and relate them to metric education. It will describe a test that was developed for metrics, comment on technical facets of constructing criterion referenced tests and present a way to maintain records for a continuous progress educational system that utilizes them.

Educational experts in areas of mathematics education, science education and educational measurement determined skills deemed important for students at the elementary level, i.e., grades 3–8. After skills had been identified, activities were constructed which were designed to develop skills in students at various grade levels. The two prior steps were basic to constructing a criterion test to measure student achievement.

The prime consideration in the development of any test is that of *validity.* Validity refers to the ability of a test to measure the purposes for which it was intended. Criterion referenced tests typically rely on *content validity.* Items are selected on a basis of being representative of the skills that have been identified. For criterion referenced tests generally the determination of the content rests with the instructor because it is generally agreed that the teacher is in the best position to know what the objectives are and whether or not the items of a test reflect the objectives. Guidelines for assuming content validity are available in several sources. Gessel,[4] 1971, Weber,[5] 1972, Candor,[6] 1974 specify the following: (1) test items should require behavior identical to the action described in the skill, (2) experts in the field should agree on the representativeness of the items, (3) responses of students should be generalizable to their responses on similar items, and (4) there must be a sufficient sample of items. A device to help assure content validity is a table of specifications which is a two dimensional outline for developing a test. The two dimensions are course content and course objectives. The table provides a test developer with a visual plan of the test and helps assure adequate emphasis on content and objectives being considered. An abbreviated specification for metric education is shown in Table I.

Table I

Table of Specifications for Metric Education

Content	OBJECTIVES 1. Knowledge	 2. Understanding	 3. Application	Total Questions
A. Weight or Mass	1	2	3	6
B. Length and Volume	2	2	3	7
C. Temperature	2	2	3	7
Total Questions	5	6	9	20

Content validity for the test included in this paper was determined in the following manner. Items were constructed that were congruent with the skills identified by experts in metric education. The content of the items paralleled the metric activities designed to develop skills in areas of mass, volume, length, and temperature. A second way used to assure content validity was through the examination of items included in the measurement section of Gessel's test. Items were selected that related to the content areas listed and were adapted for use in measuring metric skills. The Gessel test was based on a "thorough analysis of mathematics textbooks most widely used in the United States in Grades 4 through 8." Items included in it are representative of important measurement skills. The fact that many items in the test included in the current paper parallel Gessel's test is supportive of its content validity.

Finally, a third scheme used to assure content validity for the test was the provision for a section of the test to be environmentally specific. A drawback of most published tests is that students have difficulty relating situations presented in test questions to their particular situations. Test publishers have difficulty overcoming this disadvantage and frequently make little provision for the varying environments of students. They generally use test questions which are totally generalizable and which are considered appropriate for all students. Such practice is restrictive from a measurement point of view. It limits the type of situations which can be tested and relies heavily on an artificial format, i. e., paper and pencil tests. The test developed by the authors addresses the above problem by providing two sections. The first 20 questions comprise a common section consisting of multiple choice questions on metric skills deemed important for students at various grade levels. The second section of the test is an environmentally specific section which requires students to demonstrate metric skills on facets of their specific environments. The objects and situations students use in responding to the latter 30 questions in the test vary in size and quantity for different students. Answers to questions are ascertained by their teachers. However, the content of the test remains the same for all students because they are asked to respond to similar situations. In summary, content validity for the test is based on expert opinion of curricular and measurement specialists, on Gessel's test[7] which reflects the measurement content of major mathematics texts in grades 4–8 and on the inclusion of a variety of skills representative of traditional testing practice and environmentally specific activities.

One of Gronlund's[8] requirements for criterion referenced tests refers to the establishment of a standard of achievement. Two types of achievement standards are relevant to the test being examined. The first refers to the test items themselves where a decision concerning the achievement level of students at different grades had to be made. Test items were divided into three categories representing three levels of instruction.

Level 1—Grades 3 and 4
Level 2—Grades 5 and 6
Level 3—Grades 7 and 8

Items are designated in the test key as to the level for which they are appropriate. The determination of level for each of the questions was based upon information provided in the Gessel test manual and upon expert opinion.

The second achievement level needed to be addressed is that of a percent of items correct criterion, that is, the proportion of the items a student must correctly answer in order to meet the criterion. Authors cited in the reference section mention the difficulty of arriving at a set criterion level. However, the 80% level has been used in the past as an appropriate one. Such a level has been selected as the desired achievement standard for this test. Care should be exercised in applying such criteria to individual cases. A student may reach the 80% level on the total test and still be deficient in selected metric skills. A way to monitor such problems will be described in a later section of the paper.

The development of a complex rationale for constructing and analyzing criterion referenced tests has not occurred. Quantitative procedures are in a primitive stage and, while efforts to describe a system of analysis for criterion referenced tests have been attempted, none has acquired a widespread acceptance. Some of these efforts have included the adoption of classical test theory to the criterion referenced situation. In the main, such applications have not proven successful. Item discrimination, for example, refers to the ability of individual test items to distinguish between those students who scored high on the test criterion and those who have not. It has usually been computed using a point biserial or biserial correlation coefficient.

In classical test theory it is desirable to have the index to be high and positive. For criterion tests it is not important that the index be high. In certain instances it is appropriate that they are low because the achievement level for all students being tested on a particular skill may be low.

Similarly, classical reliability theory has its drawbacks when applied to criterion referenced tests. Among other factors high reliability is dependent upon variability of test scores and the difficulty level of test items. Test experts from classical schools would suggest, if maximum reliability is to be attained, that difficulty levels average in the .40–.60 range and that scores of individual students span the entire range of the test. This is not the situation for criterion referenced tests where ultimately a high percentage of the students may reach the criterion.

In spite of the above discussion, reliability coefficients are frequently computed for criterion referenced tests. Such indices have been computed for the first twenty items of the test included in this paper. In a study by Esham[9] a K. R. 20 index of .67 was reported for 134 sixth and seventh grade students. Such an index is low when compared to those of most standardized achievement tests and it is difficult to pass judgment on the coefficient in the absence of other information. For comparative purposes KR-20 coefficients for the same group on two sections of 20 items each of the McFee[10] test were .51 for the general proficiency subtest and .64 for the intuitive subtest. Other studies by Esham produced KR-20 coefficients of .64 and .67 for 117 fifth grade students and 88 sixth grade students respectively for the test described in this paper.

The final section of this paper will consider the use of criterion referenced tests for purposes of diagnosis of student deficiencies in metrics and for purposes of planning remedial instruction. For the sake of clarity of explanation a simplified example will be used rather than one based on an actual test situation. Let us assume that a criterion referenced test on metric material has been developed according to content described in Table I. That is, one question of the test relates to knowledge about weight or mass (category 1A), two questions of the test relate to knowledge about length and volume (category 1B), etc. The test length is twenty questions. Let us assume the test was administered to ten students in the class and that results were plotted on a student/class analysis chart shown in Table II.

The numbers in the first row of Table II represent the questions on the test. The symbols in the second row relate to cells in the table of specifications, Table I. The positioning of question numbers above the table of specifications referent symbols means that a given test question was meant to measure a skill indicated in the table of specifications. Subsequent rows record the achievement of individuals on specific test items, or content. A plus (+) means the student answered the question correctly; a minus (−) means he did not. At the end of the rows the achievement of each student is recorded; at the bottom of

Table II

Student/Class Analysis Chart

Question		1	2	3	4	5	6	7	8	9	10	11	12	13	14	15	16	17	18	19	20	Student Percent
Table of Specifications Referent		1A	1B	1B	1C	1C	2A	2A	2B	2B	2C	2C	3A	3A	3A	3B	3B	3B	3C	3C	3C	Mastery
	Jim Dun	+	+	+	+	+	+	+	+	+	+	+	+	−	−	−	−	−	−	−	−	60
	Sal Bee	+	+	+	−	+	+	−	−	−	+	+	−	−	+	−	−	−	+	−	−	45
S	Mary Fore	+	+	+	+	−	+	+	+	−	+	+	−	−	−	−	−	+	−	−	−	50
T	Bill Gum	+	+	+	+	+	+	+	−	+	+	+	−	−	+	−	−	−	−	−	−	55
U	Jack John	+	+	+	+	+	+	+	−	+	+	−	+	+	−	−	+	−	+	−	−	65
D	Henry Fell	+	+	+	+	+	+	+	−	−	−	−	−	+	+	+	−	−	−	−	−	50
E	Jo Fin	+	+	+	−	+	+	+	+	−	−	−	−	+	+	+	+	−	−	+	−	60
N	Joy Gire	+	+	−	+	+	+	+	−	−	−	−	+	+	+	−	−	−	−	−	−	45
T	Roy Bacci	+	+	+	+	+	+	+	+	−	+	+	−	+	+	+	+	+	+	−	−	80
S	Bob Hand	+	+	+	+	+	+	+	+	+	+	+	+	+	+	+	+	+	+	+	+	100
Subject Percent Mastery		100	100	90	80	90	100	90	50	40	70	60	40	60	70	40	40	30	40	20	10	

each column the percent of achievement of the class on each skill or objective is given.

The above data can help the teacher diagnose problem areas of the class and of individual students and assist in the planning of instruction. For example, let us assume a criterion of 80% mastery is desired. It appears that many students have not achieved the criterion. Achievement of knowledge skills (1A through 1C) has been reasonably good, for the class, as a group, has achieved 80% or above on each of them. However, achievement of application skills (3A through 3C) is poor where the class failed to achieve above 70% on any. It appears that additional instruction in the areas of application of material about metric measurement is appropriate for all members of the class, except Bob Hand.

Analysis of skills requiring understanding of metric material is a bit more complicated. Columnar totals for items 2A suggest that the class is performing satisfactorily in understanding weight or mass.

Achievement in the 2C category is below acceptable. Three students, Fell, Fin and Gire have not attained the criterion. Grouping these students and emphasizing skills measured by the test items would be appropriate. Achievement in the 2B category, understanding length and volume, is well below the criterion. Examination of the chart does not yield a pattern for planning instruction. Possibly the material should be taught again to the entire group. From the above discussion the utility of criterion testing for assisting in instruction can readily be surmised.

In this paper facets of criterion referenced tests and their applicability to metric education were discussed. Particular emphasis was given to the nature of such tests, their content validity, and their use in diagnosing pupil weaknesses and planning instruction. A fifty-item test developed by the authors of the paper was discussed. Other equally important facets of criterion referenced testing were not treated, including technical facets of item construction and the specification of metric skills and development of instructional objectives from which test items are written.

It is felt that criterion referenced testing has an important role to play in many types of classroom instruction and that teachers should consider their use as they attempt to develop metric skills in their students. Assuredly, criterion referenced testing will probably not sufficiently meet all the evaluation needs of instructional programs for metrics or any instructional program but they do have unique applicability for classroom use.

CHAPTER NOTES

1. R. Glaser and A. Nitko, "Measurement in Learning and Instruction," in *Educational Measurement,* ed. R. L. Thorndike (Washington, D.C.: American Council on Education, 1971).

2. N. E. Gronlund, *Preparing Criterion-Referenced Tests for Classroom Instruction* (New York: The MacMillan Co., 1973).

3. Harold Schoen et al., *Educational Metrics Kit: Activities and Criterion Referenced Test* (Blacksburg, Virginia: Educational Metrics Corp., 1974).

4. John Gessel, *Prescriptive Mathematics Inventory* (Monterey, California Test Bureaus: McGraw-Hill, 1971).

5. L. J. Weber and S. R. Lucas, "Evaluating Student Progress" in *The Individual and His Education, Second Yearbook of the American Vocational Association,* ed. Alfred H. Krebs (Washington, D.C.: American Vocational Association, 1972).

6. Cathie Candor, *A Model Process for Designing Criterion Referenced Placement Tests* (Charleston, West Virginia: Kanawha County Schools).

7. Gessel, *Prescriptive Mathematics Inventory.*

8. Gronlund, *Preparing Criterion-Referenced Tests for Classroom Instruction.*

9. Maurice Esham, *A Pilot Study in Methods of Metric Education* (Blacksburg, Virginia: Virginia Polytechnic Institute & State University, College of Education, 1974).

10. E. E. McFee, "The Relative Merits of Two Methodologies of Teaching the Metric System to Seventh Grade Science Students" (Doctoral diss., Indiana University, 1967).

APPENDIX
CRITERION REFERENCED TEST IN METRICS MEASUREMENT

CRITERION REFERENCED TEST

Part 1: Common Section

In this section of the test students attempt multiple choice questions about measurement. Their success will be dependent on whether or not they have mastered certain metric skills. Questions were developed for three levels of instruction:

Level 1--Grades 3 and 4
Level 2--Grades 5 and 6
Level 3--Grades 7 and 8

Correct answers to questions and the levels to which individual questions are appropriate are listed on the Key which follows the test. In order that students be considered competent they should be able to answer correctly 80% of the items listed at their level.

---

1. A meter is equal to

a. 10 centimeters
b. 100 centimeters
c. 1000 millimeters
d. both b and c

2. A kilogram is equal to

a. 100 grams
b. 1000 grams
c. 100 centimeters
d. 1000 centimeters

3. The temperature at which water freezes is

a. 0°C
b. 32°C
c. 100°C
d. 212°C

4. The temperature at which water boils is

a. 0°C
b. 32°C
c. 100°C
d. 212°C

5. What is the distance in centimeters from point A to point B?

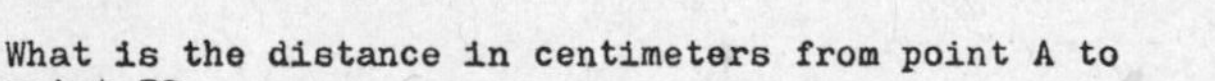

a. 1.2
b. 12
c. 120
d. 1200

6. What is the preciseness of the ruler below?

a. 1 centimeter
b. ½ centimeter
c. 1 millimeter
d. 8 centimeters

7. What is the length of the pencil?

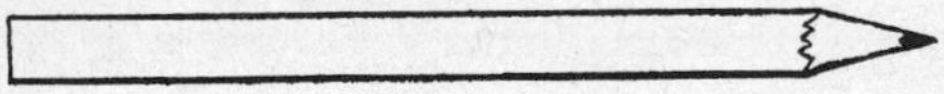

a. 1 meter
b. 1 kilometer
c. 1 centimeter
d. 10 centimeters

8. 2 centimeters and 40 millimeters equal ____ centimeters

a. 240
b. 24
c. 6
d. 60

9. Add the following lengths:

1 meter + 20 centimeters + 40 millimeters

a. 1.24 centimeters
b. 1 meter plus 240 centimeters
c. 1.24 meters
d. 124 millimeters

10. Subtract the following lengths:

10.762 meters - 3.824 meters

a. 6.938 meters
b. 7.948 meters
c. 6.948 centimeters
d. 7.948 centimeters

11. Subtract the following lengths:

147.0612 meters - 146.9887 meters

a. 1.0725 meters
b. .0725 centimeters
c. 72.5 millimeters
d. 725 millimeters

12. What is the temperature of the thermometer below?

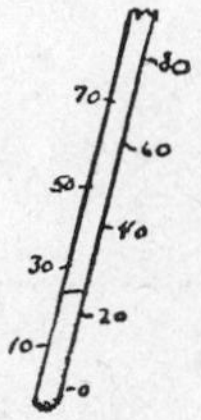

a. 20 C
b. 30 C
c. 25 C
d. 85 C

13. What is the perimeter of the following figure?

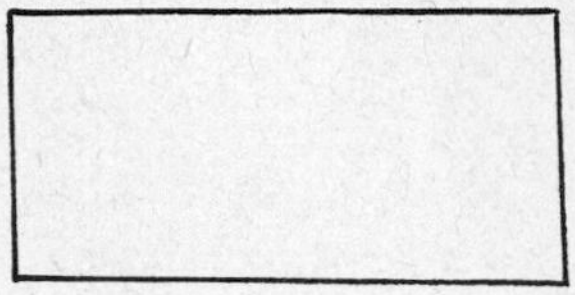

a. 18 centimeters
b. 9 centimeters
c. 12 centimeters
d. 6 centimeters

14. What is the perimeter of the following figure?

a. 7 centimeters
b. 140 millimeters
c. 14 meters
d. 1.4 meters

15. What is the area of the following figure?

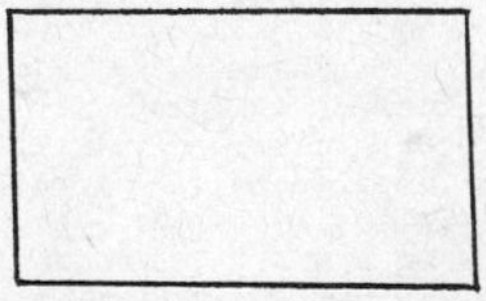

a. 8 square centimeters
b. 16 square centimeters
c. 15 square centimeters
d. 15 square millimeters

16. What is the volume of the box below?

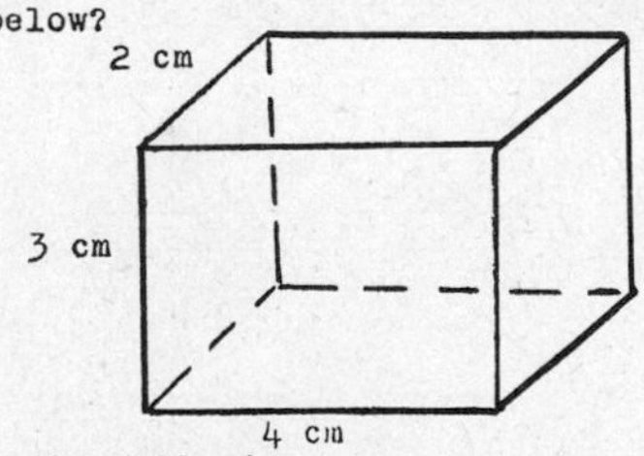

a. 7 cubic centimeters
b. 12 cubic centimeters
c. 14 cubic centimeters
d. 24 cubic centimeters

For Questions 17 through 20 refer to the following information regarding the weight of two students, Tom and Sally.

TOM SALLY

17. How much does Tom weigh?

a. 40 kilograms
b. 40.5 kilograms
c. 45 kilograms
d. 50 kilograms

18. How much does Sally weigh?

a. 28 kilograms
b. 35 kilograms
c. 36 kilograms
d. 36.5 kilograms

19. What is the combined weight of Sally and Tom?

a. 68 kilograms
b. 75 kilograms
c. 80 kilograms
d. 81 kilograms

20. What is the difference in weight between Tom and Sally?

a. 4 kilograms
b. 5 kilograms
c. 9 kilograms
d. 14 kilograms

Part 2: Environmentally Specific Section

In this section of the test students are asked to perform measurements on facets of their particular environments. Since the objects they encounter will differ in size and quantity from one group of students to another, it will be necessary for each teacher to ascertain the correct answers to the test questions. Also, it may not be possible to administer all questions to a group of students because test questions relate to things which may not be available to students. If such is the case, omit the items or devise questions using similar situations.

---

1. Place a Celsius thermometer on your desk. What is the temperature of your classroom? ________

2-4: Obtain three glasses of water according to specifications listed below. Record the temperature of the water in each glass.

2. Regular water from the tap. ________

3. Lukewarm water to wash your hands. ________

4. Water from the drinking fountain. ________

5-7: Complete the activities below and record the information.

5. Record the temperature of your classroom. ________

6. Record the temperature out of doors. ________

7. What is the difference in temperature between the classroom and out of doors? ________

8-10: Measure the top of your desk.

8. How many centimeters long is it? ________

9. How many centimeters wide is it? ________

10. What is its area in square centimeters? ________

11-13: Measure the top of your teacher's desk.

11. How many centimeters long is it? ______
12. How many centimeters wide is it? ______
13. What is its area in square centimeters? ______

14-17: Perform the activities requested and record the information.

14. Measure the length of your classroom. ______ m
15. Measure the width of your classroom. ______ m
16. What is the perimeter of your classroom? ______ m
17. What is the area of your classroom? ______ sq.m

18-20: Measure the baseball diamond at your school. Record the following information.

18. The distance from home plate to first base. ______ m
19. The distance from first base to second base. ______ m
20. The distance from home plate to second base. ______ m

21-23: Perform the following activities and record the information requested.

21. Your weight in kilograms. ______
22. The weight of two of your classmates in kilograms.

    Name ______________ ______

    Name ______________ ______
23. The average weight in kilograms of you and your two classmates. ______

24-27: Obtain six bricks and ten books. Perform the following tasks and record the information.

24. The weight of the bricks in kilograms. ______
25. The weight of the books in kilograms. ______
26. The total weight of the books and bricks. ______ kg.
27. The difference in weight between the books and bricks. ______ kg.

28-30: Weigh the following items and record the information.

28. A gallon plastic milk carton filled with water. ______ kg.
29. A sack filled with sand or dirt. ______ kg.
30. The tallest person in your room. ______ kg.

CRITERION REFERENCED TEST

Answer Key and Level Key

Part 1:

Item	Answer	Level
1	d	1/2/3
2	b	1/2/3
3	a	1/2/3
4	c	1/2/3
5	b	1/2/3
6	c	2/3
7	d	1/2/3
8	c	1/2/3
9	c	2/3
10	a	1/2/3
11	c	2/3
12	c	1/2/3
13	a	1/2/3
14	b	2/3
15	c	2/3
16	d	3
17	c	1/2/3
18	c	1/2/3
19	d	1/2/3
20	c	1/2/3

Part 2:

Item	Level
1	1/2/3
2	1/2/3
3	1/2/3
4	1/2/3
5	1/2/3
6	1/2/3
7	1/2/3
8	1/2/3
9	1/2/3
10	2/3
11	1/2/3
12	1/2/3
13	2/3
14	1/2/3
15	1/2/3
16	1/2/3
17	2/3
18	2/3
19	2/3
20	2/3
21	1/2/3
22	1/2/3
23	2/3
24	1/2/3
25	1/2/3
26	1/2/3
27	1/2/3
28	1/2/3
29	1/2/3
30	1/2/3

TOTAL IMMERSION COMPARED WITH BI-LINGUALISM METHODS USED TO TEACH THE METRIC SYSTEM

Jim F. Bassett

The metric system is the language of weights and measures based on the number ten. Over a century ago, in 1866, the metric system was legalized in the United States. Since that time there have been attempts to convert the United States, the only major power in the world not on the metric system, to the more logical system of meters, liters and grams. Recently a bill before the United States Congress was defeated which would have put our country on a scheduled change-over to the metric system. Despite this congressional action, our country is flowing in the direction of the metric system. Several major industries have set up committees and advisory boards to direct their company's conversion and the education of their employees. Market products are being labeled with both English and metric units of weight. Some road signs are displaying a new term—the kilometer. Conversion seems inevitable in the United States, and with this change there is a need to examine the process of how to convert people to a new measuring system. The intent of this study was, using elementary school children, to look at two methods of conversion and to determine if one method is more advantageous than the other.

STATEMENT OF THE PROBLEM

The problem in this study was to compare the effectiveness of the total immersion and the bi-lingualism (conversion) methods of teaching the metric system to a point where the individual is comfortable and can efficiently use meters, centi-

---

Jim F. Bassett, Ed. D. Instructor of Elementary Mathematics Methods and Supervisor of Student Teachers, Sam Houston State University.

meters, kilometers and decimeters. The study dealt only with linear units of measurement.

The total immersion method is an attempt to educate an individual through complete exposure to the metric units. There is no reference made to any English terms and no comparison of English and metric units. In the bi-lingualism (conversion) method the student trying to learn the metric system can use the more familiar English system to aid him in converting. This method may be compared to a student learning a foreign language in his native country—the student has a dual system, a crutch, or a familiar setting to lean on.

MAJOR ASSUMPTIONS

The following assumptions were made in conducting the study:

1. The children being used in the study had not been exposed previously to the metric system or had very limited exposure to it.
2. The students were familiar with the process of measuring.
3. Both programs have the same general objectives but the only major variable was the teaching method.
4. The presence of an instructor who was not the regularly assigned teacher did not significantly affect the outcome of the study.
5. The written post-evaluation is a reliable instrument in comparing the growth of the two groups of students.

HYPOTHESIS

The null-hypothesis tested in this study was: there will be no significant difference between teaching the metric system by total immersion in the metric system or by the bi-lingualism (conversion) method.

DELIMITATIONS

1. The study was limited to fifth grade students at Stewart Elementary School in Huntsville, Texas.

2. The study was limited to four one-hour sessions per group.
3. The students were limited to one program of instruction, either total immersion or bi-lingualism (conversion).
4. The students were only exposed to linear units of measurement.
5. The students had had very limited experience with the use of decimals, a concept used in the metric system.

PROCEDURE

The study was initiated with the selection of students. The participating students were members of the fifth grade at Stewart Elementary School, Huntsville, Texas. They were functioning in a mid-level mathematics class at the time of the study. There were fifty-six (56) students in all. The students had had limited or no previous exposure to the metric system. The students were randomly divided into two groups (Group A and Group B) of twenty-eight students each. Group A was the control group and Group B was the experimental group.

Instruction began with Group A. During this five-day period, Group B resumed their regular mathematics schedule with another fifth grade teacher. The students in Group A were given a pre-test to determine their knowledge of the metric system. The pre-test was followed by several different lessons and activities (see Appendix I following this chapter) designed to teach students to use and to think in the linear units of measurement in the metric system by comparing them with familiar units in the English system. On the fifth day the students were given a written post-test.

Group B began their study of the metric system the following week, with the same pre-test. Group A resumed its regular mathematics schedule with a fifth grade teacher. The lessons and activities that followed were designed to use only linear metric units (see Appendix II following this chapter). No reference was made to the English system. Group B ended with the same post-test as Group A.

RESULTS AND CONCLUSIONS

A t-test was employed to compare the means of the two groups on the post-test scores. The t-ratio was so small that the difference in the means of the two groups (1.75 points) would be due to chance more than five times out of every one hundred. Since this difference was considered not to be significant, it was not possible to reject the null hypothesis. The pre- and post-test means of the two groups are presented in Table I.

Table 1

Pre- and Post-tests Means of the Control and Experimental Groups

Group	Pre-test Mean (20 possible points)	Post-test Mean (41 possible points)
A (Control)	2.89	28.67
P (Experimental)	3.71	30.42

While the null hypothesis was not rejected, it was felt that under different circumstances (a longer period of instruction and better evaluation techniques), it might be possible to reject the null hypothesis.

Through personal observation and subjective evaluation, it was concluded that students using the total immersion method seemed to enjoy learning the metric system more than students using the bi-lingualism method. This was evidenced by a more cooperative attitude, more individual participation, and generally more enthusiasm in learning the metric system. It was therefore concluded that, in order to teach a child to "think metric," total immersion is the better of the two methods.

APPENDIX I
OBJECTIVES FOR GROUP A (CONTROL GROUP)

Day 1

1. To use a non-standardized unit of measurement, the human hand, to develop an awareness and a need for a standardized unit of measurement.
2. To use a standardized unit of measurement, a black strip of construction paper representing a decimeter, without knowledge that it represents a decimeter, to measure familiar items in the classroom.
3. To have students conclude that a longer unit of measurement than a decimeter is needed; and to construct one by taping the black strips of paper together.
4. To have students conclude that the number ten (10) is a good base for a measurement system.
5. To have students conclude that a smaller unit of measurement is needed and to cut the black strips of paper into smaller units of measurement.
6. To label the units of measurement in the metric system used for linear measurement (millimeter, centimeter, decimeter, meter, kilometer).
7. To compare the units of measurement in the English system used for linear measurement (inch, foot, yard, mile) with the units of measurement in the metric system used for linear measurement (millimeter, centimeter, decimeter, meter, kilometer).
8. To observe a chart of the units of measurement in the metric system as compared with the units of measurement in the English system.

Day 2 - Objectives

1. To review the linear units of measurement in the metric system, their relationship to each other, and the relationship to the linear units of measurement in the English system.
2. To learn the abbreviations for the metric units of measurement.

3. To each make a meter stick and divide it into decimeters and then into centimeters.
4. On a worksheet first guess the length of familiar items in English units of measurement and then in metric units of measurement; then measure the items on the worksheet with an English system tape measure and the meter stick the students made.

Day 3 - Objectives

1. To measure and compare each student's body measurements in inches and centimeters and record these measurements on a worksheet.
2. To convert from one unit of measurement to another unit of measurement.

Day 4 - Objectives

1. To review converting from one unit of measurement to another unit of measurement.
2. To measure lines on a worksheet in both English and metric units of measurement and to compare the relationship of English and metric units of measurement.
3. To develop a conversion formula from inches to centimeters. Example: 1 inch = 2 1/2 centimeters
4. To decide the best unit of measurement to use when measuring different items.
5. To estimate the length of items in centimeters, then measure the items in centimeters.

Day 5 - Objectives

1. To review the metric units of measurement, their relationship to each other, and their abbreviations.
2. To administer the post-test.

APPENDIX II
OBJECTIVES FOR GROUP B (EXPERIMENTAL GROUP)

Day 1

1. To administer a pre-test to determine the student's prior knowledge of the metric system.
2. To use a non-standardized unit of measurement, the human hand, to develop an awareness and a need for a standardized unit of measurement.
3. To discuss items that could be used as a standard unit of measurement.

Day 2

1. To use a standard unit, the decimeter, without knowledge that it represents a decimeter, to measure familiar items in the classroom.
2. To have students conclude that a longer unit of measurement is needed and to construct one by taping ten black (decimeter) strips together.
3. To also conclude that there is a need for a smaller unit of measurement by measuring items less than one decimeter in length.
4. To make a smaller unit of measurement by dividing the decimeter strip into ten equal parts
5. To label each unit as a meter, a decimeter or a centimeter.
6. To make a chart of the metric terms, their relationships to one another, and their abbreviations.

Day 3

1. To make a meter stick and divide it into decimeters and centimeters.
2. To have each student use his meter stick to measure the length of lines in centimeters.
3. To review the metric terms and their relationship to one another by viewing a filmstrip.

4. To estimate, without the use of any measuring device, the length of certain items in a specific metric unit.

Day 4

1. To have the students use their meter sticks to measure parts of their own body and to record the measurements in centimeters.
2. To estimate the length of familiar classroom items in metric units.
3. To have the students evaluate their estimations by actually measuring the items with their meter sticks and recording the measurements.
4. To convert larger metric units into smaller units.
5. To draw lines of a certain length in both centimeters and decimeters.
6. To work independently in the work packet to provide practice and application in working with the metric system.

Day 5

1. To have the students convert smaller metric units into larger metric units.
2. To allow students to apply the new ideas they have learned by working in their work packets.
3. To evaluate the student's progress.

CLOSING REMARKS

Gomer Pound

It is reflective of today's conditions that after two and one-half days of total immersion in metrication, you surface to find a *Pound* bidding you Godspeed; for despite your earnest efforts—and no one can question your zeal—this country has not officially changed to SI. A pound is still understood better by more people in this country than a kilogram.

At the close of last year's conference I stated that forces had already been set into motion in this nation that would inevitably compel a change to metric measurement: that Congress notwithstanding, the die was cast. After another year congressional action now, as then, is pending. But there is no cause to be disheartened. I am reminded that it took one hundred and fifty-five years before Congress offically adopted a national anthem, following more than a half century of unofficial, and on occasion official, use of Francis Scott Key's famous poem as our national song. And Congress has yet to accede to repeated requests to establish a code for performance of the anthem. Be patient.

Actually, congressional fiat, though unquestionably helpful, is not the ultimate answer to adoption of SI in this country. After all, prohibition and universal daylight saving time did not endure. Acceptance by the American public is, as you well know, the key. Congress will follow suit.

The sensible approach to accelerating the build-up of momentum that will eventually cause Congress to acknowledge SI is through massive educational engagement. That is why your preoccupation here during these three days takes on added importance.

Gomer Pound Dean, Division of Extension and Public Service, University of Southern Mississippi.

I spoke of "massive" educational engagement. By this I meant a program that cuts through every stratum of society, embracing all age groups. Some colleges and universities can be of special service to metric education in this regard through their commitment to continuing education.

We in the division of Extension and Public Service at the University of Southern Mississippi take satisfaction from the role we have been able to play in metric education. Working closely with the department of science education, we have conducted workshops and classes over a widespread area in addition to coordinating conferences like this. At present a course in metric education is being readied to be made available through our department of independent study. We have arranged for rental of the film depicting the country of Zambia's experiences with metric education that attracted enthusiastic notice at last year's conference.

Continuing education units possess experience in working with students of all ages, from a variety of backgrounds, in diverse situations. They have for many years gone to where the action is: schools, factories, businesses, storefronts, wherever the need for instruction might be. Their network of delivery systems and contacts is established and proved. Professionals in continuing education can be of invaluable assistance to a program such as that required to effect the conceptual, attitudinal, yes, emotional, changes that will eventuate in universal adoption of SI.

It is axiomatic that change does not create conflict: resistance to change creates conflict. Therefore, when you devise educational strategies be sensitive to circumambient issues. I recall, for example, that last spring the crafts unions lobbied to delay congressional action on the metric bill because of the issue of whether their members would incur the costs of new tools. Perhaps an educational thrust in their direction could have forestalled their interference with passage of the bill.

In thinking about what I might say to you today it occurred to me—and I say this only partly facetiously—that Congress might solve problems of recession by mandating SI. The number of jobs that could be created due to retooling, retraining, reprinting, and repurchasing, could stimulate a sluggish economy if the transition took place during a relatively short period. Perhaps the thought can be cultivated for transmission to your representatives in congress.

Regardless of the merit or feasibility of my last suggestion, do be alert to opportunities for educating specific populations of every description. You cannot concentrate your efforts in the schoolroom alone and accomplish your purpose, unless you are reconciled to a long, long, wait. We will do well to heed Adlai Stevenson's aphorism, "Change is inevitable, but change for the better is a full time job."

Thank you for coming. Call us whenever we can be of assistance. Our division's motto is "We deliver". We will make every effort to help. I bid you Godspeed.